The Astronomer Jules Janssen

Astrophysics and Space Science Library

For further volumes:
http://www.springer.com/series/5664

Françoise Launay

The Astronomer Jules Janssen

A Globetrotter of Celestial Physics

Translated by Storm Dunlop

Springer

Françoise Launay
Observatoire de Paris
Paris, France
Francoise.Launay@obspm.fr

ISBN 978-1-4614-0696-9 e-ISBN 978-1-4614-0697-6
DOI 10.1007/978-1-4614-0697-6
Springer New York Dordrecht Heidelberg London

Library of Congress Control Number: 2011936386

Translation from the French language edition: *Un globe-trotter de la physique céleste: L'astronome Jules Janssen* by Françoise Launay, © Vuibert 2008. All rights reserved

Cover photo: "Statue of Janssen at Meudon" courtesy of the author

Printed on acid-free paper

Springer is part of Springer Science+Business Media (www.springer.com)

Fig. 1 *Antoinette Janssen in 1907*
© Paris Observatory Library

This book is dedicated to Antoinette Janssen, who did so much for her father's memory.

A senior nursing officer in the Association des dames françaises [Association of French Ladies] during the First World War, the sole daughter of Jules Janssen did, in effect, take the initiative in ensuring the publication of her father's *Œuvres scientifiques*.[1] This work collects together the scientific articles that she herself and Henri Dehérain, the curator of the Library of the Institut de France, judged to be the most significant. Henri Dehérain was the son of Pierre-Paul Dehérain, a lecturer at the Muséum, the friend that Janssen knew in 1852 when they were students together at the Sorbonne, and who was elected to the Académie des sciences in 1888, during his presidency.

The two volumes were published, respectively, in 1929 and 1930, after Antoinette's death.

[1] Janssen, Jules, *Œuvres scientifiques*, recueillies et publiées par Henri Dehérain, Paris, Société d'Éditions Géographiques, Maritimes et Coloniales, vol. I (1929) and vol. II (1930).

Antoinette Janssen also left two significant legacies:

- The first, to the Library of the Institut de France, of all the correspondence received by the family as well as her father's notebooks, ordered chronologically, like diaries.
- The second, to the Conservatoire national des arts et métiers (CNAM), of three instruments designed by Janssen: the 1862 spectroscope, the 1870 aeronautical compass, and the 1874 photographic revolver, the recognized predecessor of the cine camera.

The collection of scientific works saves a lot of research in scattered publications. The instruments represent a major tribute to Janssen as an inventor, who was definitely more of an experimenter and engineer than a theoretician. The notebooks and the correspondence[2] are an extraordinary mine of information, because, apart from the some 500 letters exchanged by the Janssen husband and wife couple, the collection includes more than a thousand others sent to Janssen by more than 160 correspondents. The latter consist not just of people from the scientific world, but also the literary and artistic worlds, and even politicians.

[2] An annotated edition of this correspondence, augmented by letters from other sources, is in preparation, *cf.* Launay, Françoise, "Jules Janssen's in and out correspondence," in *100 years of observational astronomy and astrophysics, Homage to Miklos Konkoly Thege (1842–1916)*, C. Sterken and J. Hearnshaw, eds., Brussels (2001), 159–68.

Preface to the French Edition

Jules Janssen. This D'Artagnan of science, this bard of the Sun, and this audacious master builder, merits far more than just a statue on the public terrace at Meudon, at the opposite end to the observatory gates: a statue subject to the ravages of time. Janssen is actually the founder of astrophysics in France. After confirmation of the triumph of Newtonian theory, and both the subtle and powerful developments of it made by Clairaut, Lagrange, and Laplace, and after Le Verrier and the discovery of Neptune, the need arose to study the nature of the celestial bodies and of that of the

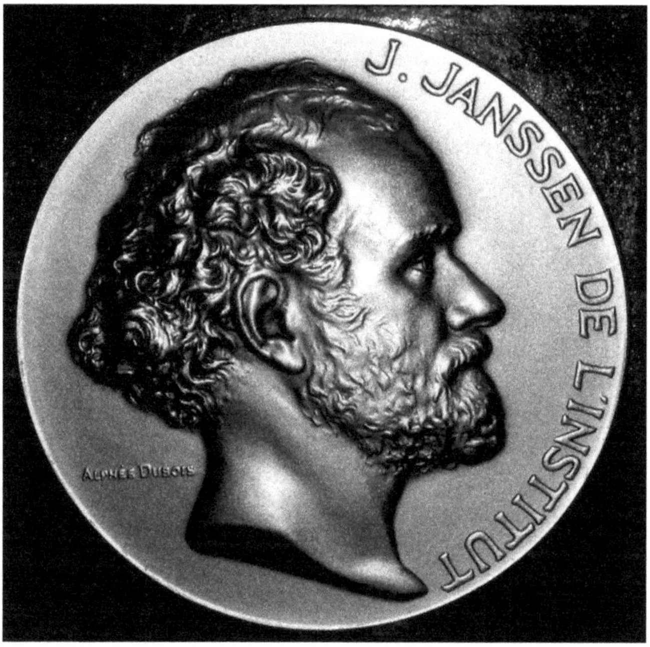

Fig. 8 *The Janssen Medal*
© Author's collection

Sun in particular. What a long delay there was, in our country, where celestial mechanics developed remarkably, but in splendid isolation, relative to the astrophysical work carried out by Anglo-Saxon and German astronomers. There was Herschel with the infrared and the "island universes"; Fraunhofer and the solar spectrum; Wolf and Schwabe and the observation of solar activity...

In this book, which, let us be in no doubt, will become a source of reference, all the aspects of Janssen's rich personality have been covered by Françoise Launay, his attentive and erudite biographer. A physicist, an inventor and instrument builder, crazy about the Sun and travel, Janssen was a sort of "self-made man," who had acquired most of his great knowledge, practically alone, guided by his energy and curiosity. Naturally, a mathematician (he followed the courses given by Cauchy and Le Verrier), he tried (without success) to be recruited by the tetchy director of the Paris Observatory. But luck accompanied his employment as the tutor that he became, and brought him in contact with the future great traveler and geographer, Alfred Grandidier. From this began his great travels around the globe, under various pretexts. Initially, this was an effort directed toward geophysics; in Peru, Janssen measured the magnetic field and determined the magnetic equator. The spirited young man had no hesitation: in preparing for his trips, he made contact with the most competent members of the Académie. He built up, if I may put it this way, a really useful address book... He threw himself into physics: his thesis discussed the propagation of heat. But in what medium? Not in optical laboratories. He became a physiologist: the medium was the eye, which was not damaged by the near infrared (then called "thermal") radiation. It was this that formed the very personal subject of Janssen's thesis, a thesis in which there was no intervention from the hand of any master. Janssen's first invention was, in effect, Janssen himself.

Like all inventors, before arriving at the correct solution, he tried everything. The analysis of the spectrum of sunlight revealed the role of absorption in the Earth's atmosphere. So Janssen's research followed two paths: the Earth's atmosphere and the Sun. And mastery of a technique, spectroscopy, that would enable him to make rapid progress toward those two ends.

Studying the Earth's atmosphere and the Sun encouraged travel, even demanded it. Nothing could have appealed more to the enthusiastic temperament of our physicist. And was it not on a mountain that one had the best of the Sun without being bothered by the Earth's atmosphere? But was it not also there that the purest spectra were available? A high mountain, then. Switzerland, the Alps, Etna, the Himalaya... Janssen was restless. And not just to observe the Sun! A map of the globe devoted to his journeys would be criss-crossed by an extraordinary number of erratic round trips and covered in small flags "Jules Janssen was here." Just look at Janssen's biography, so minutely picked out by Françoise Launay in her "biographical landmarks"! It is absolutely astonishing, given that in Janssen's day there were no TGVs or Airbus 320s... Among this almost Brownian motion, we might pick out two – no, sorry, two groups – of journeys that are particularly memorable.

One was the trip to Japan in 1874 to observe the transit of Venus in front of the Sun. In 1882, recognizing that celestial mechanics was not his strong point, Janssen did not take part in the expeditions for the second transit. The transits of Venus provided plenty of "grist to the mill" for the specialists in fundamental astronomy. But it was not our astrophysicist's "cup of tea," which he readily recognized. Admittedly, he invented the famous photographic revolver to record the precise instants of the moments of contact during the course of the transits. And it was a success. But it was more of a technical advance rather than an astronomical event.

From a scientific point of view, observation of the Sun during total eclipses, which had to be carried out at the far corners of the Earth, was of far greater significance.

It is true that in the history of astronomy in France, we find a few tentative attempts in solar physics. Lalande was interested in sunspots, which were measured by some of his students, Flaugergues, for example, and whose observations were used to establish the cycle of solar activity. But it was Janssen who really paved the way. Captivated for life by his first encounters with the Sun, Janssen would never relinquish it; he monitored it from the four corners of the Earth. In 1867, it was the eclipse at Trani, the very first. In 1868, the voyage to India, the total eclipse of the Sun, and his results were not only remarkable, but were noticed. This was the beginning of Janssen's great notoriety. Then there came the saga of the total eclipses of 1870 (Oran), 1871 (India), and 1875 (Siam). There were others: in 1883, it was Caroline Island, in the Pacific, a long way away…

It reached the point where Henriette frequently complained of the solitude in which her traveler of a husband left her, caught, poor woman, by having to care for a small daughter and her mother-in-law ("between an old woman and a child"): "There are men who leave their wives for mistresses; you do it for journeys!" He finally took her with him to India in 1871, where she carried out a very sympathetic sociological study of the Todas. Henriette had good reason to want to follow her husband. In many respects, she could have gone before him. But the times were not favorable for wives, who were expected to confine themselves to baby clothes and casseroles.

Françoise Launay accurately describes the progress that Janssen made in an understanding of the Sun. There is no point in recalling here the new knowledge of the spectrum of the corona. There is no point in recalling here the discovery of the solar nature of the prominences, nor of their observation outside eclipses (anticipating the spectroheliograph of Deslandres, his unpleasant successor). Again there is no point in recalling his magnificent work on the disk of the Sun, achieved much later at Meudon, and the success of recording granulation photographically. These major advances are known to all astronomers and marked decisive stages in knowledge of the solar chromosphere and photosphere.

These successive discoveries, achieved thanks to an exceptional combination of a rare instrumental inventiveness and a tireless obstinacy in observing, meant that Janssen was among the world's leaders in astrophysics. Rivalries arose at this level, notably with Secchi, who classified stars according to their color (Janssen had also dipped his toe slightly in that water), but friendships were also established that were more durable. With Norman Lockyer, his rival over the eclipse of 1868; and with William Huggins, the stellar spectrographer. And in France itself, there were connections that were both pleasant and very useful. Ministers, such as Victor Duruy,

Léon Gambetta, and Jules Simon or later, Jules Ferry and Léon Say, the rich senator and astronomer Raphaël Bischoffsheim and the deputy and engineer Ernest Cézanne. And then there were the figures that were well-known in scientific circles: Camille Flammarion, Prince Roland Bonaparte, and Gustave Eiffel. There was the romantic painter Jean-Jacques Henner, who specialized in chlorotic young girls lying, more-or-less extended, on rather indistinct couches, and who also painted portraits of Jules … and many others. We can well imagine Janssen, a key figure of the turn of the century, having pride of place on some sort of flower-bedecked platform, surrounded by all those bearded and triumphant celebrities: Léon Gambetta, Jules Ferry, Charles Gounod, Louis Pasteur, Sully-Prudhomme, Auguste Bartholdi, Camille Saint-Saëns…

But the man was not indifferent to the struggles of his century. During the siege of Paris in 1870, he left the capital in a free balloon, entrusted by Jules Simon, the Minister of Education, in Paris with a mission to Gambetta, Minister of the Interior and of War, who had fled to Tours. We may wonder if Janssen would have attempted this hazardous trip if there had not been a solar eclipse at the end of it. But nevertheless he did it! It seems quite clear that Janssen was a patriot and republican, and that he was open to new ideas.

Basking in the glow of his success, Janssen could attack what was, for him, even more significant than the solar prominences or photospheric granulation; more than the telluric lines of water vapor or optics of the eye; more than the hectic travels of an eclipse observer or of the transits of Venus in front of the Sun – his great work: construction of a great astrophysical observatory. Construction? That's saying a lot! It was first essential to create the observatory on paper, *in the abstract*. That was not easy; Janssen the traveler was not often there to defend his project, and the decision took a long time to materialize from the paperwork. As Hervé Faye said in his enthusiastic report: "Because physical astronomy can no longer be merged with astronomical mechanics, let us give it a separate establishment: the two sciences will thus develop in parallel, without interfering with one another, and using differing expertise." We sense the reluctance of older astronomers, and probably of Le Verrier himself. Among astronomers, Faye was thus the strongest supporter for the creation of the observatory, but Henriette, always the Parisienne, doubtless played a considerable part. She valued the idea of this establishment, the hoped-for stability of which would ensure her comfort from the more constant presence of her rather volatile husband.

The observatory and the directorial pair were finally established at Meudon. It was in the charred ruins of the "Château neuf" at Meudon, which had been converted into a powder magazine during the war, which is never a guarantee of a favorable outcome, that Janssen installed a dome, a large refractor (the largest in Europe, and rivaling the one at Lick – the Yerkes instrument did not yet exist), a reflector, and a spectroscopy laboratory, as well as astronomers of some status: Henri Perrotin, Gaston Millochau, Henri Deslandres, and Albert Nodon.

It remained for this far-sighted astronomer to set up an observatory on a high mountain. This was the foundation of the observatory on Mont Blanc, the astronomical counterpart of the Refuge Vallot, which was devoted to geophysics.

Our indefatigable traveler, still suffering from his limp, took part in the climb on a ladder carried by porters. And this courageous sexagenarian “with his mass of hair and snow-white beard,” observed the Sun with a spectrograph and showed that the oxygen lines were of terrestrial origin.

Although the dangers posed by the ice on the peaks forced the authorities to demolish the observatory on Mont Blanc (in 1909), that at Meudon remains one of the pinnacles of astronomy in the world today.

Facing the observatory terrace and the great dome, recently refurbished with a new covering of copper, there is the statue of Jules Janssen. On the base of the monument, three bas-reliefs (unfortunately partially destroyed; their restoration is a duty that ought to be undertaken by the community at large) illustrate the astronomer’s life. The one that remains represents the balloon trip during the Siege, another evoked the expedition to the East to observe the eclipse of the Sun in 1868, and the third the dangerous ascent of Mont Blanc: all symbols of the indefatigable and prolific activity of the founder of French astrophysics, Jules Janssen.

Paris, France Jean-Claude Pecker

Preface to the English Edition

Jules Janssen has long been a hero of mine, so it is with pleasure that I introduce the fine biography of this important inventor and astronomer written by Françoise Launay of the Paris Observatory. Ms. Launay has delved into family papers, contemporary newspaper articles and reports, and even stone monuments in Meudon and in Père-Lachaise cemetery in Paris. My comments supplement the earlier preface by the distinguished astrophysicist Jean-Claude Pecker.

Janssen had several triumphs in the early spectroscopy of the mid-nineteenth century. He was able to determine the absorption of water vapor in the air, matching with laboratory spectra and showing the dependency of the absorption on the length of the slanted passage of sunlight through Earth's atmosphere. One can trace back a lot of today's studies of global climate change to this work. He even took spectra of the erupting Santorini volcano.

Janssen has been known to astronomers especially for his pioneering work in taking a spectrograph to the total solar eclipse in India in 1868, at almost 7 min in duration about the longest possible in duration, and for his discoveries there. In particular, Janssen and the contemporary Englishman Norman Lockyer are credited with the near-simultaneous discovery that the solar prominences can be observed even outside of eclipse by looking at the red spectral line of hydrogen now known as H-alpha. We learn in this biography of just how Janssen undertook the expedition to India and about what the various scientists who observed the eclipse actually saw. As might be expected from any excellent biography, the question of who saw what when is more complicated than the traditional naive view. We see that though Janssen saw the yellow spectral emission line soon said to be from "helium," since it existed only on the Sun, he did not single it out or identify it as special. (Launay's section title is "Janssen and the helium that he did not discover.") We see how the question of priority of the discovery of this method of observing prominences, which indeed led to the discovery of the chromosphere itself surrounding the sun (a level named by Lockyer) was resolved in a satisfactory fashion for posterity by sharing the credit. Indeed, the two profiles appear together on a bronze medal struck subsequently in France.

The spectroscopic eclipse work led to the following year's eclipse observations, though it was other astronomers than Janssen who first saw the so-called coronal green line, for decades said to be from "coronium" in analogy to the "helium" spectral line discovered in 1868. It was this coronal green line along with other, subsequently discovered coronal lines, that eventually led to the identification of the corona as million-degree gas, an idea that would have astounded Janssen.

I had not realized that Janssen was so self-motivated and so-self taught, having been brought to his maturity outside the scientific and academic environment of the French "grand schools" that provided the elite. It was not until his 40s that Janssen arranged independently with the minister to observe the key 1868 solar eclipse, bypassing channels and the official Paris Observatory expedition set up by Le Verrier, whose fame for predicting the position of what turned out to be the planet Neptune had led him to the directorship.

When Janssen decided to observe the 1870 total solar eclipse, he would not let a little thing like Paris being surrounded by the Prussian army stop him! He arranged to escape from Paris in a balloon. His liaison with Lockyer was sufficiently well established that the English arranged a safe passage for him directly with Bismarck. But we learn from Launay that Janssen was carrying a secret political message for Gambetta, whom he went to see before continuing to Algeria for the eclipse. Janssen just did not think it ethical to use Bismarck's safe passage in those circumstances.

Janssen went on to observe several other eclipses, such as the 1871 eclipse in India, at which he concluded that the corona was a self-luminous solar atmosphere, and the 1875 eclipse in Siam, the one that led to the King of Siam (famous for his liaison with his governess, Anna, made known to US audiences through the musical *The King and I*) not only to attend the eclipse but also to die soon after from the malaria contracted from a mosquito at the eclipse site. Those interested in the roles of women in nineteenth-century science will be able to read about the writings of Mme Janssen, who for her description of the natives she met during the 1871 expedition, "in addition to the respect of great men, the article brought Henriette 15 centimes a line."

Launay has discovered the previously unknown or unappreciated background of inventors in Janssen's family, which explains some of the scientist's greatest triumphs. One of the rarest predictable and observed events in astronomy is transits of Venus. They occur in pairs separated by 8 years, with intervals of 105.5 or 121.5 years between the pairs. Nobody alive on Earth had seen a transit of Venus before the one in 2004. Now we eagerly await the June 6, 2012, transit of Venus, which will be visible at sunset in the continental USA (and entirely in Hawaii and Alaska) and at sunrise in Europe (except for Portugal and western Spain) and as far east as India and Western Australia.

For the 1874 transit, Janssen wanted to be able to time accurately the contacts of Venus's limb and the Sun's, something that was foiled the previous century by the "black-drop effect," in which a ligament joined Venus's dark silhouette with that of space outside the solar limb (edge) and took a minute to stretch and pop. The uncertainty prevented timing of the requisite accuracy to meet Edmond Halley's requirements for measuring the distance between the Earth and the Sun and hence the scale of the entire solar system.

Janssen thus worked out an annular daguerreotype with an intricate system of rotating wheels that allowed 48 exposures to be taken at 1.5-s intervals. This

"Janssen revolver" (known in England as simply as a "Janssen") was the forerunner of multiple imaging, beginning the path that led through Muybridge's study of locomotion to the movie cameras of Thomas Alva Edison and the Lumière brothers.

My wife and I were fortunate this spring to see an example of Janssen's revolver on display at the Espace Pierre-Gilles de Gennes (a 1991 French Nobel laureate in Physics), an exhibition space at the location in Paris, l'École supérieure de physique et de chimie industrielles de la Ville de Paris, where Marie Curie had separated out her radium from piles of ore a few decades after Janssen's work and a century before the present. The display cases contained one of Janssen's revolvers, with the gears showing. Alas, the only one annular daguerreotype made of the transit cannot be now found, though three plates from preliminary tests do exist. A few months earlier, my wife and I had seen a different version of the revolver at Sydney Observatory in Australia, a remnant of one of the British expeditions. Janssen himself had observed the 1874 transit of Venus in Japan, while contenting himself with providing apparatus for the 1882 event.

Janssen went on to perfect his high-speed photography enough to capture the best images of solar granulation that were then available.

Janssen left a monument to his achievements in the Meudon Observatory, founded in the late nineteenth century as an independent institution in a suburb of Paris. The observatory includes the "Grande Lunette," the third largest refracting telescope in the world. I was fortunate that Audouin Dollfus, a great French planetary astronomer of this century, arranged for several of us to observe Mars at its 1988 opposition, when it was closest to Earth and therefore largest and brightest in its 26-year period (a synodic period mimicked in our times by the launches of spacecraft to Mars at that interval). The observatory incorporated a chateau made available to Janssen for the project, all on a series of terraces overlooking Paris at a distance. I cannot wait to return to Meudon to see again the monument to Janssen at the end of one of the terraces, no doubt the one from which my family and I once saw the most magnificent fireworks ever to celebrate the 14th of July.

The Meudon Observatory is now jointly run with the Paris Observatory, with the former's monument to Janssen a long way from the statue of Jean-Dominique Cassini, the first of a dynasty of four that left their mark on the latter's, in the museum of that seventeenth-century edifice. It is fitting that Launay's office should now be in the ancient, seventeenth-century observatory building, across the hall from a ceiling mural showing beautiful, naked Venus in her chariot meeting Apollo's chariot as she traverses the sky.

As we come to the planned high-cadence photography of the 2012 transit of Venus, celebrate new high-speed-camera results about insects and flowers, enjoy slow-motion views of sports on television, and follow eruptions on the Sun with the "3D Sun" app on our iPads and iPhones (http://3dsun.org), it is fitting to ponder the contributions from Jules Janssen that have led to these aspects of our modern world. Françoise Launay's book brings us to the appreciation of Janssen's oeuvre that he deserves.

Williamstown, MA Jay M. Pasachoff

Acknowledgements

The idea of writing this book would undoubtedly not have come to me if, some 15 years ago, doctor Pierre Amalric, the great historian of ophthalmology, since deceased, whom I had met at the Pic du Midi Observatory, and who wanted to write an article on Jules Janssen's links with his specialty, had not asked me: "Do you have a portrait of this man at Meudon?"

The positive reply was given to me by Audouin Dollfus, in whose laboratory I started my career as an engineer, and whom I had the great pleasure of meeting again in the group known as the "Patrimoine scientifique de l'observatoire de Meudon" *["Scientific heritage of the Meudon Observatory"]*, which he had been running for some time. An astronomer and aeronaut keen on new breakthroughs, just as Janssen had been, Audouin Dollfus introduced me to the forceful personality of the founder of our observatory, who, I realized, carried out laboratory spectroscopy at the precise location where the large spectrograph was installed, and for which I was responsible technically.

It was my colleague Simone Dumont who told me, slightly later, about the existence, in the Library of the Institut de France, of a rich Janssen collection, and who provided me with a detailed inventory.

I would like to heartily thank all three for their decisive involvement.

I also benefited from a marvelous welcome at all the institutions where I consulted documents concerning Janssen, and express my gratitude to those who enabled me to have access to the various collections involved: Mireille Pastoureau, Annie Chassagne, and Fabienne Queyroux at the Library of the Institut de France; Christiane Demeulenaere-Douyère and Florence Greffe at the Académie des sciences Archives; Mireille Lamarque at the Institut de France Archives; Marie-Renée Cazabon at the Collège de France Archives; Patrick Fuentes and Jacques Pernet at the Library of the Flammarion Observatory of the Société astronomique de France at Juvisy; Peter Hingley at thc Royal Astronomical Society's Library in London; Adam Perkins at the Cambridge University Library; George Wilkins at the Norman Lockyer Observatory at Sidmouth; Rodolphe Rapetti and Claire Bessède at the Jean-Jacques Henner Museum, Quentin Bajac at the Musée d'Orsay, André Gunthert at the Société française de photographie; Marie-Sophie Corcy at the Conservatoire national des art et métiers; and Laurence Bobis, Annie Accary, Josette Alexandre, and Nathalie Reymonet in the Paris and Meudon libraries of the Paris Observatory.

It was also particularly pleasant to establish fruitful contact with Françoise Andlauer, Jean-Pierre de Beaumarchais, Bertrand Dufort, Anne Guigan-Léauté, Pierre Modot, and Nicole Worms de Romilly, all of whom I warmly thank for having agreed to plunge into family archives on my behalf, and to share unpublished documents and information about Janssen and his family.

My deepest thanks go to my colleagues and friends who have offered me their support during the course of my research: Maurice Benharrous, Marta Braun, Danielle Briot, Henri Chamoux, Yvon Georgelin, Peter Hingley, Christophe Huber, Georges Laroche, Jean-Yves Roncin, René Sigrist, and Philippe Véron, as well as to those who have given my manuscript a careful and critical reading: François Biraud, Christine Etienne, Évelyne Mérigot, Gualtiero Olivieri, and Michel Toulmonde.

I am extremely grateful to Jean-Claude Pecker for having provided such a fine preface to this work; one that encourages the reader to follow Janssen on his numerous trips.

I am, finally, indebted to Laurence Bobis, head of the Paris Observatory Library, and to Éditions Vuibert, in particular to Marc Jammet and Alain Luguet, for all the confidence they were kind enough to show in me in publishing this work.

The American edition of this book would certainly have not seen the light of day without the decisive impetus given by Bill Sheehan, whom I first met in London and Paris, and subsequently "on the shores of the Moon's Mare Orientale," and who was able to find a receptive ear at Springer Science+Business Media. My thanks to Harry Blom for having taken on the venture so swiftly and to his collaborator, Jessica Fricchione, for having started the work.

Storm Dunlop carried out translation of the work. It was undoubtedly a great privilege to be able to benefit from his participation (many thanks to André Heck for having acted as intermediary), and it was a real pleasure to work with one of the most renowned specialists in this field. I would like to thank him most warmly here for his judicious remarks, his attention to my comments, and for his rewarding suggestions.

Sincere thanks to Jay Pasachoff for having been so kind as to put aside Venus for long enough to write an additional preface, which proves that Jules Janssen still has fine trans-Atlantic admirers.

Contents

Chapter 1
Foreword

Fig. 1.1 *Jules Janssen. Photograph on glass plate.*
© Paris Observatory Library

It is the 31 October 1920, and we are on the wide public terrace on the Meudon estate, in front of the "château neuf" of Louis, the Grand Dauphin, heir to Louis XIV, who died before his father. It was Jules Janssen who transformed the building into an observatory at the beginning of the 1880s.

F. Launay, *The Astronomer Jules Janssen: A Globetrotter of Celestial Physics*, Astrophysics and Space Science Library 380, DOI 10.1007/978-1-4614-0697-6_1,

It is fine, and is the day of the unveiling of the statue of Janssen, who died in 1907. The project dated back to 1911. The committee for the monument, formed under the chairmanship of Henri Poincaré, the mathematician, member of the Académie des sciences and of the Académie française, originally included twenty-two members, all members of the Institut de France. It was finally to gather together sixty-one prominent subscribers, including the President of the Académie des sciences, Armand Gautier; the Vice-President, Gabriel Lippmann, winner of the Nobel Prize for Physics; Benjamin Baillaud, Director of the Paris Observatory; Henri Deslandres, Director of the Meudon Observatory; Edmond Perrier, Director of the Natural History Museum; Doctor Roux, Director of the Pasteur Institute; and eighteen foreign scientists: two British, two German, one Italian, one Spaniard, one Dutchman, one Swede, one Swiss, two Russians, six Americans, and His Highness, Prince Albert 1st of Monaco. Among the names that were well-known to the general public were those of Gustave Eiffel, Camille Flammarion, the explorer Jean Charcot, and Baron Pierre de Coubertin.

Fig. 1.2 *The Château neuf at Meudon turned into an observatory. Postcard.* © Author's collection

Fig. 1.3 *Statue of Janssen at Meudon. Janssen statue inauguration ceremony.* © Author's collection

Antoinette Janssen was perhaps the sole representative of the family at the ceremony: her mother Henriette had died in 1913, cousin Eugène Janssen in 1908, and cousin Henri Le Roux, director of departmental affairs for the Seine prefecture, who was a member of the committee, in 1914. Only the daughter of the last, Anne, and her husband, the former deputy Raoul Bompard, as well as the cousin Albert Dufort, an engineer at the Centrale [Grande École of Engineering], might have been among the hundreds of people present. What is certain is that a large number of important people were present at the official unveiling, even if a fair number of the members of the "Tout-Paris" set to which he once belonged, had disappeared. The war had, in fact, prevented the project from being completed as quickly as might have been desired. Missing from the distinguished foreigners were, in particular, Oskar

Backlund, Director of the principal Russian observatory,[1] Sir David Gill, Royal Astronomer at the Cape and, above all, Sir Norman Lockyer, Janssen's great English friend, who had died two months earlier. Others, who could not attend made their feelings known by some significant token. This was the case with the American George Hale who sent "flowers enclosed in the folds of a flag covered in stars".[2]

Guillaume Bigourdan represented the Committee, the Académie des sciences, and the national Paris Observatory, which, as he said, not without a certain sense of humor, "warmly shares in the joy of its young brother, Meudon".

Henri Deslandres represented the Bureau des longitudes and, of course, Meudon Observatory, then still independent of the Paris Observatory, to which it would be linked only in 1927,[3] and of which he had been the director since the death of Janssen.

His Highness Prince Roland Bonaparte represented the Geographical Society, the French Alpine Club, the French Photographic Society, the Belgian Photographic Association, and the Aero-Club of France.

Camille Flammarion naturally represented the Sociéte astronomique de France, which he founded in 1887, and which Janssen joined in 1891.

Lieutenant-Colonel Paul Renard, the brother of the famous Colonel Charles Renard, who was also a major figure at Meudon and who died in 1905, represented the Permanent International Commission for Aeronautics, and the French aeronautical societies.

Other societies to which Janssen had belonged, such as the French Association for the Advancement of Science, the Meteorological Society, the Société philomathique, the Pigeon-fanciers Society, the Société d'hygiène, the French Red Cross, and the group known as the Société La Marmite undoubtedly sent representatives.

Alfred Dubuisson, the accounts clerk was there on behalf of Janssen's former colleagues at the Meudon Observatory.

Finally, Julien Luchaire, principal private secretary to the Minister for Education and Fine Arts, André Honnorat, was there to accept the monument in the name of the state.

The contemporary account tells us that the monument consists of "a statue in white Carrera marble, measuring 2.70 m high, sculpted by [*the architect*] Hippolyte Lefebvre (grand prix de Rome), and a base in Chauvigny stone, measuring 3.20 m high, 2.55 m wide, and the same in depth, built by M. [*Jacques*] Debat-Ponsan, architect (prix de Rome). Three faces of the pedestal were decorated by bas-reliefs sculptured by H. Lefebvre".[4]

When the covering fell away, we are told that it "revealed the superb silhouette of the scientist and thinker, reproduced with marvelous skill, and a perfect likeness". With his right wrist resting on a globe and an optical prism in his hand,

---

[1] Pulkova Observatory.

[2] Bigourdan, Guillaume, in *Inauguration de la statue de Jules Janssen*, Paris, Gauthier-Villars, 1920, 9—14, p. 10.

[3] Decree dated 05 October 1926.

[4] *Inauguration de la statue de Jules Janssen*, p.3.

Fig. 1.4 *Janssen's ascent of Mont Blanc: crossing a crevasse. Postcard.* © Author's collection

Janssen holds several sheets in his other hand. Is this the orator, reading one of the talks for which he was known, or is it the thinker, who is offering passers-by the fruits of his thoughts and which he has set down on paper? Rather, we are prepared to bet that Janssen is looking at the notebook of drawings of the solar prominences that he made in India in 1868, using a prism to isolate the red light from them, and that Lefebvre was inspired by a photograph of Janssen, standing with just such a document in his hand.

One of the bas-reliefs on the base represents that eclipse, whilst of the other two, one recalls Janssen's escape from besieged Paris in the *Le Volta* balloon in 1870, and the other where "the scientist, lying on a makeshift structure"[5] crosses a crevasse in the Mont-Blanc massif, which recalls the foundation of his observatory at the summit in 1893.

All the speakers got on with their fine speeches. It is true that there was plenty for them to say, because Janssen's career was more than just rich. After all, are there not some ten inventions to his account, as well as multiple laboratory experiments, and various observations of which there are six observations of total solar eclipses carried out in all four corners of the Earth, hundreds of talks and lectures, and the creation of no less than two observatories?

After mention of all these accomplishments, "the persons present dispersed, taking with them positive memories of this celebration of Science, which had taken

[5] *Le Petit Journal*, 31 October 1920, p. 6.

Fig. 1.5 *Janssen as a benefactor of humanity. One of 84 portraits of "benefactors of humanity" enclosed within chocolate wrappers by the firm of Guérin-Boutron. The text reads: "Janssen, French scientist, born 1824, the author of numerous scientific works, went to the summit of Mont Blanc to continue his research into light, radiation, etc."*
© Author's collection

place before an intellectual elite, in the superb setting of the Meudon grounds, in the peaceful calm of a beautiful autumn day, fully in harmony with a life that had been dedicated to pure and disinterested research into Truth".[6]

How, then, are we to explain the fact that Janssen, who was among the "five hundred contemporary celebrities" in the first album of photographs published by Félix Potin[7] in 1904, and in which there were no more than seventeen scientists, engineers and explorers,[8] should be so poorly known nowadays?

Was his name not inscribed, like that of Pasteur, on the pediment of the Science and Industry wing of the Grand Palais, on the plan[9] that the architect Gustave Rives drew up in 1896?

How could Janssen, even in his lifetime have been classified as "a benefactor of humanity" on the wrappers of bars of chocolate?

[6] *Inauguration de la statue de Jules Janssen*, p.4.

[7] Potin was a grocery company. The photographs could be found inside the wrappers of all their products.

[8] Apart from Janssen, there were Adolphe Alphand, Claude Bernard, Marcelin Berthelot, Alphonse Bertillon, Jean-Baptiste Boussingault, Paul Broca, Eugène Chevreul, Gustave Eiffel, Hervé Faye, Camille Flammarion, Charles Hermite, Ferdinand de Lesseps, Louis Pasteur, Alfred Picard, Henri Poincaré, and Louis Troost.

[9] Musée d'Orsay archives, ARO 1985-27.

How was it that this grandson of a cobbler, who had to start work early merely to live, and who graduated neither from the prestigious École Normale Supérieure nor École Polytechnique (the highest French "grandes écoles"), was able, expedition after expedition, discovery after discovery, invention after invention, to become known and recognized by the highest scientific and political authorities, to such a point that, for him, they created the institution that faces his statue: this fine *Observatoire d'Astronomie Physique de Paris, sis Parc de Meudon*, deliberately independent of the great observatory in the heart of Paris, planned for France two centuries earlier by Colbert and Louis XIV?

Chapter 2
Childhood and Education

"La nature ne livre ses secrets qu'à ceux qui ne reculent devant aucun effort pour les lui arracher."

[Nature only reveals its secrets to those who do not flinch from any effort to wrest them from it.]

Jules Janssen [1]

A Family of Inventors

Although three forenames Pierre, Jules, and César were given when his birth was registered, Jules, and only Jules, is the name that should be used for Janssen, who was born at his parents' house in Paris, 14 rue l'Évêque, on 22 February 1824, at 4 in the morning [2]. This was how he was known during his lifetime, and that is the form he used as his signature, although on most occasions he simply used the initial: J. Janssen. Jules was also to be the sole forename that was to be inscribed on his tomb. His parents were Antoine César Janssen, a musician, 42 years old, and Pauline Marie Lemoyne, his wife, aged 35. The witnesses who signed the registration of birth were Pierre Louis Puzin, a dealer in passementerie[1] of rue Saint-Denis, and Jean François Joseph Grandvalet, an upholsterer of rue de Provence. The baby was baptized the next day, at the church of Saint-Roch [3].

Let us establish at the outset that Jules Janssen's family name should be pronounced "Jeanseine" [*French pronunciation of "Jean"* + *"Seine"*], and that any attempt at a Flemish or Scandinavian pronunciation is an insult to his memory. Indeed, the words of a letter that Mme Janssen sent to her husband on 13 February 1869 are particularly relevant at this point:

> I read *[the following]* in *La Liberté* for 6 February: "A Dutch chemist, Janssen, according to *Le Moniteur Universel*, has just determined, by means of spectral analysis, the existence

[1] Trimmings of gold or silver lace, braid, beads, etc., for clothing and other objects.

F. Launay, *The Astronomer Jules Janssen: A Globetrotter of Celestial Physics*, Astrophysics and Space Science Library 380, DOI 10.1007/978-1-4614-0697-6_2,

of water vapor in the atmosphere of Mars and Saturn." Is *Le Moniteur* already suffering from the Carnival spirit? It has turned M. Janssen, our learned French physicist, Doctor of Science at Paris University, into a Dutch chemist! [4]

The usual sources regarding Janssen's ancestry are obituary notices, of which there are many, frequently inspired rather too freely by earlier biographical notes, and which generally copy one another. Only the astronomer Guillaume Bigourdan, from the Paris Observatory, tried to find out more by seeking proper sources, but he was not able to confirm all that he read, nor all that he had been told – not even, it would seem, by Janssen himself ("son and grandson of artists ..." [5]). In addition, it was undoubtedly because he did not want to cause Janssen's widow and daughter any distress that he did not mention everything he learned from the documents he consulted; proof of this is shown by his notes, which have been preserved in the archives of the Académie des sciences. The result is that when various authors claim that nothing predisposed Janssen to a scientific career because he came from a family of artists (his father is described as "a talented musician" and his maternal grandfather is said to be the architect of La Madeleine and a friend of Beaumarchais), they are rather far from the truth. For a start, they conceal the fact that his father was just an ordinary second clarinet in the orchestra at the Opéra-Comique, who "never rose to be anything other than mediocre" [6], as Fétis dares to maintain in his *Biographie universelle des musiciens!* On the other hand, they do not lay enough emphasis on the fact that Antoine César invented the roller clarinet and was probably deprived of the patent by an instrument maker. Fétis mentions a favorable report about this by the mathematician Francœur [7], and Mme Beau speaks of the "Janssen key," which "earned its inventor a silver clarinet that was presented to him by the *Société des Artistes*" [8]. In addition, something that no one knew, and which we were interested to discover, is that some of the boot makers and manufacturers of passementerie on Janssen's paternal side possessed patents. Janssen undoubtedly came from a family of inventors, and he did not disgrace them.

Moreover, when biographers have stated that his maternal grandfather, the architect Lemoyne, who admittedly won the Prix de Rome, took part in the construction of La Madeleine, they forgot to consult the corresponding file, which tells us that he only submitted one of the 82 designs in the competition. In his *Dictionnaire des architectes français*, Lance is at least as critical as Fétis: "He produced a plan for the restoration of the vaults and great doorway at the cathedral of Sens. [...] Luckily, the Revolution affected this project, which was not one of restoration, but of mutilation" [9]. As for the link with Beaumarchais, the idea should be completely dismissed: Beaumarchais did not know Lemoyne personally before engaging him as one of the two architects for his celebrated Paris mansion, and their relationship rapidly deteriorated [10].

Janssen did not, in fact, know any of his grandparents. His paternal grandfather, Christianus Janssen, a master cobbler, who died in Paris in 1797 at the age of 60, had been born in what, at the time of his death, was known as the French department of la Dyle [11], i.e., the Brussels region. Christianus, called Chrétien or Christian on various official papers, had married a Parisienne, Louise Catherine Puzin, sister of the dealer in passementerie from the rue Saint-Denis. The children born of this marriage were Marie Louise, who married a dealer in ladies' shoes, Jean Dufort (whose two sons became boot makers and whose daughter became a cobbler); and then Antoine César, the father of Jules, Antoine Vincent, a cobbler, as well as Noël Henry Janssen, a boot maker, who was a witness at the religious marriage ceremony of Jules' parents in 1818 [12].

On the maternal side, Janssen's grandparents were the architect Guillaume Paul Lemoyne and his first wife Albertine Pieters, widow of a man she married at Ghent in 1777. Lemoyne, who died 14 years before Janssen was born, was the son of a building contractor and a lady née Le Doux, who came from a family of craftsmen, of which a number were master gilders and other jewelers, and who had been resident in Paris since the seventeenth century.

As far as the family's fortunes are concerned, it is highly probable that Lemoyne's, despite his having won the Prix de Rome in 1775, and contrary to what is usually stated, was non-existent. In fact, his name only appears on the list of entrants to the competition that was opened on 20 December 1806 for the construction of the "Monument à la gloire des armées" that became La Madeleine because "he is not rich" and it was decided to reimburse, to the extent of 600 francs, his costs for detailed drawings for plans that he had sub-contracted [13]. After becoming a widower, in Paris in 1794 he married Marie Thérèse Lefevre. The son from this second marriage, Georges, was some 12 years old when his father died, and Janssen mentions in his notebooks how his parents gave a home to the second Mme Lemoyne in June 1842 until her death, which occurred in 1843.

On the Janssen side, it would seem that the financial situation was, at least originally, far better, because Antoine César had a private income from 1835, even though he was then just 54 years old.

The Bank Employee and Secondary Education

Jules' parents, who intended him to follow a career in painting [14], had decided that he should study drawing, but nevertheless did not neglect a classical education:

> He was an only son, and as such, very pampered, even more so because an accident caused by his nurse's carelessness rendered him lame *[at the age of 8]*. So, contrary to the more-or-less general usage at the time, he was not sent to a boarding school, but carried out all of his studies as an external student. [15]

His varied abilities rapidly became apparent:

> Even as a child he had a flair for observation that foreshadowed his scientific vocation, and which was accompanied by considerable dexterity. He excelled at cutting out magnificent deer-hunting scenes from paper, with huntsmen, the pack of hounds, and his daughter still preserves a small, wooden box, decorated with tacks, which he made at the age of 8, and which shows an ingenuity and patience that is extremely rare at that age. [16]

In 1840, just 1 year after Arago had described, both to the Chamber of Deputies [17] and to the Académie des sciences and the Académie des beaux-arts [18], Niepce and Daguerre's method of capturing images in a camera obscura, a photograph of

Fig. 2.1 *Jules Janssen at 16*. Enfance des célébrités contemporaines. © Bibliothèque de l'Institut de France

Jules was taken (Fig. 2.1). We know of this portrait because it featured in the Exposition de l'enfance ["Exhibition of Childhood"], held in Paris in 1901, where it was described as: "An extremely old, and well-preserved daguerreotype, which reveals the features, at 16, of M. Janssen, the illustrious astronomer, who has become so well-known from his many discoveries" [19].

It was precisely at this period, according to all the biographies, that Jules' parents lost their fortune. Although no known document gives details of exactly what happened, we may imagine that it was as the result of some bad financial investment. It is definite, however, that Janssen had to give up his studies and turn to working to earn his living. From this time onwards, he maintained highly detailed accounts. He continued to do so throughout his life and never spent even a centime unnecessarily.

The Janssens left rue des Bons-Enfants, where they had been living since 1836 for rue Rochechouart when, in October 1840, Janssen joined the office of the bank owned by the brothers Tharaud (Bertrand, Gustave, and Léopold). He remained there until 30 September 1847, exactly "7 years less 12 days" [20]. By recording it in this way in his notebooks, we must assume that he found it tedious, even though he made good use of his time, and he had a future actor at the Vaudeville Theatre, Adolphe Dupuis, as a colleague.

> Janssen told Henner[2] that Dupuis and he were employed in the same bank at the age of eighteen. Both were trying to find their true vocation. Even while dealing with the correspondence, one would recite scenes from Molière, while the other studied mathematics. The manager pulled a face at Molière, but looked on Janssen's books with respect. Each eventually followed their own path. [21]

---

[2] This is Janssen's great friend, the painter Jean-Jacques Henner (1829–1905).

In 1842, the Janssens settled in Montmartre, first in the rue Royale, then, in 1843, at 40 chaussée Clignancourt. On 1 October 1847, Janssen changed his employer; he worked for M. Boulet, whom he left in July 1848 to move to M. Lapelle's firm [22].

Janssen, who "learned arithmetic, algebra and trigonometry on his own, from Bezout, and then infinitesimal calculus from Lacroix" [23], followed the course at the Conservatoire on Sundays, and continued to study on his own, learning Latin and Greek. He gained his baccalaureate[3] of letters in January 1849, and of mathematical sciences (for which he then became eligible) in November 1850, after having followed the course given by the physicist Paul Desains at the collège Bonaparte.

Higher Education and Early Travels

In 1851, at the age of 27, Janssen began his higher-education studies at the Sorbonne where "he followed the courses given by Cauchy, Lefébure, Le Verrier, and Sturm" [24]. He could not yet afford the "Grand Tour," the trip lasting several years that at that period young people of good family undertook, but he went to London for the month of August. He obtained his licentiate[4] in mathematical sciences in July 1852, before staying in Switzerland for the next 2 months.

In his notebook for 1853 [25], while acting as supply teacher at the lycée Charlemagne, he recorded that he got in touch with Urbain Le Verrier, a graduate of the École Polytechnique and Director of the Paris Observatory, with a view to a possible collaboration. Le Verrier was known for "being as affable as a bulldog" [26], but Janssen does not specify whether he was ever taken on at the Observatory.

In October 1854, he was hired by the rich lawyer Jacques François Napoléon Grandidier to be tutor to his sons Ernest and Alfred, at the château de Fleury-Mérogis. The sons became great travelers and the latter of the two like Janssen became a member of the Académie des sciences as a geographer and navigator. Janssen obtained his licentiate in physical sciences in November 1855.

The following year, he finally undertook the major trip that he had undoubtedly been dreaming of making for a long time: Constantinople, Rhodes, Cyprus, Jerusalem, Alexandria, Cairo, Malta...

He had, however, decided to devote his life to research, and so the time had come for him to carry out an official scientific mission, which was to be the first in a long series of such undertakings.

[3] The baccalaureate (Fr. "baccalauréat") is the required qualification for university entrance.

[4] The licentiate is a degree awarded by certain European universities, intermediate between those of bachelor and master (or doctor).

Determination of the Magnetic Equator in Peru

Janssen decided to travel to South America with the Grandidier brothers “for the purpose of resolving certain questions about the physics of the globe” [27]. He intended to make magnetic observations in Peru and then in Chile, and called on the help of three members of the Institut, the physicist Babinet, and the two hydrographers Daussy and Duperrey in planning the trip. In the first demonstration of his strong will and astonishing self-assurance, on 12 October 1857 Janssen told the Minister for Public Education and Worship about his plans, and asking for a mission assignment. He wrote to the Minister: “We would therefore ask you, Monsieur le Ministre, to be so kind as to agree to our undertaking and to give it an official character, only which will guarantee its entire success” [28]. He added that he expected an answer “as rapidly as possible” because the departure was set for 19 October, that is, a week later! Janssen did not ask for money (it was undoubtedly the Grandidier family that was underwriting the finances), and on 24 October, the minister agreed to this “free” mission [29]. However, things by no means transpired as had been wished. Janssen fell seriously ill in Peru: dysentery, intermittent fevers, and hepatitis immobilized him for several months, and even his return to France had to be deferred. Treated at Lima, he was finally in a fit state to return to Paris in August 1858, without having been able to complete his scientific work.

With the Schneider Family at Le Creusot, Marriage, and His Doctoral Thesis

At the beginning of 1859, when he officially became engaged to Henriette Forestier (who remained faithful to him for 2 years, despite the snide remarks from one of her Grandvalet cousins regarding her future husband’s lack of fortune), he left for Le Creusot to work for the Schneider family. He had been engaged by Eugène Schneider, the director of the ironworks, a former Minister of Agriculture and Commerce and Deputy, as tutor to his son Henri. Janssen had to get him ready for his baccalaureate, which was, apparently, not an easy task. This was undoubtedly not Janssen’s only occupation, as he wrote to Henriette on 5 February:

> At the moment, I am overwhelmed with jobs: what with the mathematical and chemical work that I give Henri, I have my personal tasks, which I am finding very difficult because of my distance from Paris. In addition, M. Schneider is thinking of establishing a school for young engineers at Le Creusot, and has asked me to draw up a programme of courses for all the different forms of science that feature in the work of the engineer. So I am forced to consider teaching methods of the principal engineering schools in France. [30]

He was lucky enough to have at Le Creusot, not only a laboratory where he could set his pupil to carrying out experiments, but also a “physics cabinet” [31] to use for his personal research.

At the same time, he was also preparing Henri Le Roux, the son of his first cousin, Geneviève Élisa Janssen (and who was 2 years younger than Henri Schneider, having been born in 1842), for his baccalaureate.

Janssen returned to Paris in August 1859 to marry Henriette the following month. The marriage contract [32] of 17 September details the property that the future couple (Figs. 2.2 and 2.3) would bring to their joint estate. The list was not very long. The clothes, linen, personal jewelry, silverware, and furniture were estimated, respectively, at 2,645 F [*francs*] for Jules and 3,210 F for Henriette. To this should be added Henriette's books, estimated at 300 F and, for Jules, "his library, scientific instruments, curiosities and collection of weapons, the whole to a value of 5,300 F." In addition to some bonds in the Chemins de fer du Nord, du Midi, and d'Orléans, Jules had at least 2,185 F in cash, but his contribution was burdened by about 2,000 F, which represented the sums that he still had to pay on income from a state loan, to which 1,060 F had to be added that he owed to his future mother-in-law. For her part, Henriette brought five shares in the Chemins de fer du Nord with a nominal capital of 500 F, and a credit note for 500 F. The settlement expected from the estate of her grandfather, who died in 1855, would bring more administrative complications than revenues.

The civil marriage took place on 19 September 1859, and the religious ceremony on the 21st at the church of Saint-Merry, in Henriette's parish. There was then no longer question that the young woman, who, until then had given lessons to support her existence and that of her mother, should continue to do so. Instead, her husband asked her to go back to school, and learn English and science. But let there be no mistake: as neither of the newly weds had a fortune, it was affinity and love that brought about the marriage. A few days before, Jules had sent these loving and rather sensual lines to Henriette from Le Creusot:

> These bonbons will be kept and, when I arrive in Paris, a charming little hand that I know will pop them into my mouth, then rosy lips will fleetingly touch mine in a sweet kiss. A way of eating sweetmeats that makes them singularly agreeable; I recommend it to all those in love and whose heart no longer completely belongs to them. [33]

Once married, Henriette went to live with her husband at the Janssen home at 87 chaussée Clignancourt. She thus left her mother, who had been the widow of Michel Forestier since 1849. Henriette's paternal family originated in Clermont-Ferrand, where Michel Forestier was a haulage contractor. He had married Uranie Véronique Thierry in Paris, and it was there that the daughters were born, Caroline in 1823, and Henriette on 12 November 1828. Their younger brother, Eugène, was born in 1830. It was at Clermont-Ferrand, where she remained until 1852, that Henriette had her schooling, at a boarding school. Henriette's sister married the merchant Léonard Sénèque Blémont in 1842, and her brother, who was in business in Le Havre and then in the United States (where he died in 1891), did not marry until 1874.

Through her mother, Mme Forestier was a cousin not just of the Grandvalets, friends of the Janssens, but also of the mother of Eugène Labiche.[5] For Jules and

---

[5] A well-known French dramatist (1815–1888), particularly noted for his comic farces and vaudeville pieces.

Fig. 2.2 *Henriette Janssen, née Forestier. Photograph by Étienne Carjat.* © Archives of the Académie des sciences of the Institut de France

Fig. 2.3 *Jules Janssen. Photograph by H.J. Whitlock, Birmingham.* © Archives of the Académie des sciences of the Institut de France

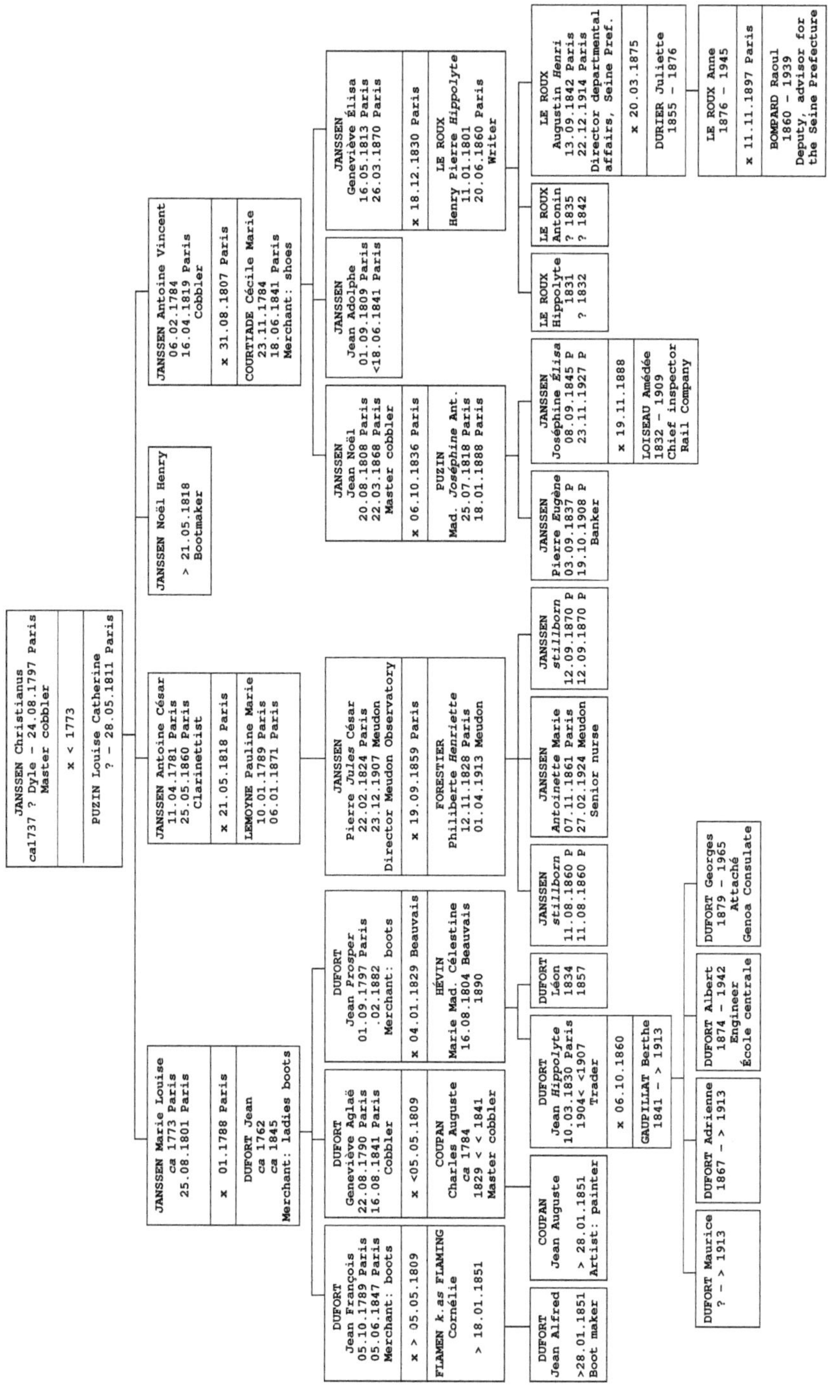
JANSSEN Christianus
ca1737 ? Dyle - 24.08.1797 Paris
Master cobbler
x < 1773
PUZIN Louise Catherine
? - 28.05.1811 Paris
JANSSEN Marie Louise
ca 1773 Paris
25.08.1801 Paris
x 01.1788 Paris
DUFORT Jean
ca 1762
ca 1845
Merchant: ladies boots
JANSSEN Antoine César
11.04.1781 Paris
25.05.1860 Paris
Clarinettist
x 21.05.1818 Paris
LEMOYNE Pauline Marie
10.01.1789 Paris
06.01.1871 Paris
JANSSEN Noël Henry
> 21.05.1818
Bootmaker
JANSSEN Antoine Vincent
06.02.1784
16.04.1819 Paris
Cobbler
x 31.08.1807 Paris
COURTIADE Cécile Marie
23.11.1784
18.06.1841 Paris
Merchant: shoes
DUFORT
Jean François
05.10.1789 Paris
05.06.1847 Paris
Merchant: boots
x > 05.05.1809
FLAMEN k.as FLAMING
Cornélie
> 18.01.1851
DUFORT
Geneviève Aglaë
22.08.1790 Paris
16.08.1841 Paris
Cobbler
x <05.05.1809
COUPAN
Charles Auguste
ca 1784
1829 < < 1841
Master cobbler
DUFORT
Jean Prosper
01.09.1797 Paris
.02.1882
Merchant: boots
x 04.01.1829 Beauvais
HÉVIN
Marie Mad. Célestine
16.08.1804 Beauvais
1890
JANSSEN
Pierre Jules César
22.02.1824 Paris
23.12.1907 Meudon
Director Meudon Observatory
x 19.09.1859 Paris
FORESTIER
Philiberte Henriette
12.11.1828 Paris
01.04.1913 Meudon
JANSSEN
Jean Noël
20.08.1808 Paris
22.03.1868 Paris
Master cobbler
x 06.10.1836 Paris
PUZIN
Mad. Joséphine Ant.
25.07.1818 Paris
18.01.1888 Paris
JANSSEN
Jean Adolphe
01.09.1809 Paris
<18.06.1841 Paris
JANSSEN
Geneviève Élisa
16.05.1813 Paris
26.03.1870 Paris
x 18.12.1830 Paris
LE ROUX
Henry Pierre Hippolyte
11.01.1801
20.06.1860 Paris
Writer
DUFORT
Jean Alfred
>28.01.1851
Boot maker
COUPAN
Jean Auguste
> 28.01.1851
Artist: painter
DUFORT
Jean Hippolyte
10.03.1830 Paris
1904< <1907
Trader
x 06.10.1860
GAUPILLAT Berthe
1841 - > 1913
DUFORT
Léon
1834
1857
JANSSEN
stillborn
11.08.1860 P
11.08.1860 P
JANSSEN
Antoinette Marie
07.11.1861 Paris
27.02.1924 Meudon
Senior nurse
JANSSEN
stillborn
12.09.1870 P
12.09.1870 P
JANSSEN
Pierre Eugène
03.09.1837 P
19.10.1908 P
Banker
JANSSEN
Joséphine Élisa
08.09.1845 P
23.11.1927 P
x 19.11.1888
LOISEAU Amédée
1832 - 1909
Chief inspector
Rail Company
LE ROUX
Hippolyte
1831
? 1832
LE ROUX
Antonin
? 1835
? 1842
LE ROUX
Augustin Henri
13.09.1842 Paris
22.12.1914 Paris
Director departmental
affairs, Seine Pref.
x 20.03.1875
DURIER Juliette
1855 - 1876
LE ROUX Anne
1876 - 1945
x 11.11.1897 Paris
BOMPARD Raoul
1860 - 1939
Deputy, advisor for
the Seine Prefecture
DUFORT Maurice
? - > 1913
DUFORT Adrienne
1867 - > 1913
DUFORT Albert
1874 - 1942
Engineer
École centrale
DUFORT Georges
1879 - 1965
Attaché
Genoa Consulate

Fig. 2.4

Henriette this meant not only that they occasionally received theatre tickets, but also that they were invited by the Labiches to their property at Souvigny, in Sologne.

Because Henri Schneider was not ready to take his baccalaureate at the time of the marriage, Janssen had to stay at Le Creusot. Henriette accompanied him, but the couple were at Paris at the time of the death of Janssen's father, on 25 May 1860. Henriette, who was pregnant, did not return to Le Creusot with her husband in June, and this first long separation weighed heavily on her. She remained in Paris with her mother-in-law, while Jules continued to teach his pupils on the one hand, and to work on his thesis on the other. His research, inspired by observations that he was able to make among the workers in the workshops of the ironworks, related to "The absorption of dark heat radiation[6] within the components of the eye"[7]: the cornea, lens, vitreous humor, and the eye as a whole. Janssen therefore devoted himself to working on spectroscopic photometry, and the conclusion that he drew from his observations and his many measurements was that the near infrared, which is absorbed by aqueous media, does not cause any damage to the retinas of higher animals. These results were not revolutionary, but the research was carried out very conscientiously, and Janssen utilized a considerable number of eyes from recently slaughtered animals (bullocks, calves, sheep, pigs, and even chickens, in his preliminary studies of the transparency of the cornea! [34]) to carry out his experiments.

Henriette expected to return to Le Creusot after the birth of her "baby" [35] (which was how she spoke of it, using the English word), once the thesis had been defended.

Unfortunately, it was not to be. Although Janssen received his science doctorate at Paris on 17 August, the little girl who was born 6 days earlier was "a child without life" [36]. Henriette had to recover from the ordeal. So she remained in Paris while Jules left, and had to deal with the first scientific controversy: a person called Cima maintained that he had carried out the same experiments earlier, and that his conclusions, published in August 1859, were very different. Janssen defended himself vigorously, and things ended to his advantage:

> However, the small battle I have had with this Austrian from Turin also ended with complete victory: Monsieur Cima who, what's more, is by no means an Austrian, neither by birth nor in behavior, has written me a very generous letter saying that he has considered my dissertation with great care and that he fully recognizes that I am right, and that he accepts my results in preference to his own. He authorizes me to inform my supervisors at the Institut. I am writing to him to ask him if, by means of a letter to the Académie, he would confirm his support for my results. So that, my dear Henriette, is what you need to get printed in Le Moniteur.[8] [37]

We learn, in passing, that the matter had gone quite far. Henriette was delighted with the outcome: "As for M. Cima, I feel that he has acted as a generous and sensible person; he ought to be French" [38].

---

[6] This is what we would now call the near infrared.

[7] This is the subject of the thesis that he was to defend.

[8] The journal in question is *Le Moniteur scientifique*. Here, as subsequently, any passages that are underlined are as they appeared in the original quotations.

Janssen, who was 36 years old, was working without a break, and dreamt of better days, but he still had to wait a long time for them to arrive:

> As for the life I live here, I am so greatly occupied with my two pupils who have to be pushed forward at a great rate that I hardly have time to do a bit of work for myself: So my second piece for the Archives Médicales has not yet been finished. I am not at all happy to find that I am always engrossed in working for my livelihood, while I feel that there are so many things to do. If only I could live, free from all such material considerations, just for my two mistresses: the one to whom I have given my heart, and the other to whom I am constantly attracted. [39]

Janssen finally left Le Creusot at the end of October 1860. Returning to Paris, he carried on with his work on physiological optics and collaborated with the doctor Eugène Follin, a renowned ophthalmologist at the Paris hospital La Salpêtrière, who became his friend. It was not long before a joint publication [40] appeared on a new form of ophthalmoscope. This was Janssen's first design of instrument.

In March 1861, the Janssens moved, and settled with Janssen's mother in rue Labat, at what was then number 21.[9] It was there, still in Montmartre, that little Antoinette was born on 7 November.

References

1. Janssen, Jules, "Note sur l'éclipse du 19 août dernier", *Comptes rendus hebdomadaires des séances de l'Académie des sciences (CRAS)*, **105** (22 August 1887), 365–6.
2. From Jules Janssen's birth certificate, Paris archives.
3. Parish register of the Roman catholic church of Saint-Roch.
4. Letter from Henriette to Jules Janssen, 13 February 1869, Bibliothèque de l'Institut de France (BIF) Ms 4134-265.
5. Janssen, Jules, "Conférence sur la photographie faite à l'issue du congrès international de photographie, le 17 août 1889", in *Œuvres scientifiques*, vol. II, 153–66, p. 159.
6. Fétis, François-Joseph, *Biographie universelle des musiciens*, Paris 1874, Tome 4, p. 426.
7. Francœur, Louis Benjamin, *Bulletin de la Société d'encouragement*, **21** (1822), 40–5.
8. Mme Henri Beau, "Jules Janssen", in *Enfance des célébrités contemporaines*, Paris, Revue L'Enfant, 1913, 23–9, p.23.
9. Lance, Adolphe, *Dictionnaire des architectes français*, Paris, 1872, Tome 2, p. 54.
10. Personal communication from Jean-Pierre de Beaumarchais to the author, 1996.
11. From the register of deaths, Paris archives.
12. Paris archives.
13. Archives nationales (AN) F^{21} 576.
14. Janssen, Jules, "Conférence sur la photographie faite à l'issue du congrès international de photographie, le 17 août 1889", in *Œuvres scientifiques*, vol. II, 153–66, p. 159.
15. Mme Henri Beau, "Jules Janssen", p. 24.
16. *Ibid.*
17. Sessions of 15 June and 3 July 1839.
18. Joint session of the two Academies of 19 August 1839.
19. Da Cunha, A., "L'exposition de l'enfance", *La Nature*, No. 1463, 08 June 1901, 63–4.

[9] Number 21 later became number 33.

20. Jules Janssen, notebook, BIF Ms 4126–2.
21. Durand-Gréville, Émile, *Entretiens de J.-J. Henner*, Paris, Lemerre, 1925, p. 262.
22. Jules Janssen, notebook, BIF Ms 4126–2.
23. Mme Henri Beau, "Jules Janssen", p. 27.
24. *Ibid.*
25. Jules Janssen, notebook, BIF Ms 4126–2.
26. Touchatout [Bienvenu, Léon Charles], *Le trombinoscope*, No.62, Paris, 1873.
27. Letter from the Minister for Public Education and Worship to Jules Janssen, 24 October 1857, AN F^{17} 2972.
28. Letter from Jules Janssen to the Minister for Public Education and Worship, 12 October 1857, AN F^{17} 2972.
29. Letter from the Minister for Public Education and Worship to Jules Janssen, 24 October 1857, AN F^{17} 2972.
30. Letter from Jules Janssen to Henriette Forestier, 05 February 1859, BIF Ms 4133-2.
31. Letter from Jules Janssen to Henriette Forestier, 10 April 1859, BIF Ms 4133-6.
32. Marriage contract dated 17 September 1859, AN, MC/ET/CVII/911.
33. Letter from Jules Janssen to Henriette Forestier, 1859, BIF Ms 4133-22.
34. Janssen, Jules, "Note sur une propriété de l'œil relativement à la chaleur rayonnante", sealed envelope deposited at the Académie des sciences, 12 September 1859, in *Œuvres scientifiques*, vol. 1, 1–5.
35. Letter from Henriette to Jules Janssen, 27 June 1860, BIF Ms 4134-4
36. Death certificate, Paris archives.
37. Letter from Jules Janssen to Henriette Forestier, 14 September 1860, BIF Ms 4133-31.
38. Letter from Henriette to Jules Janssen, 27 June 1860, BIF Ms 4134-10.
39. Letter from Jules Janssen to Henriette Forestier, 14 September 1860, BIF Ms 4133-32.
40. Janssen, Jules and Follin, Eugène, "Considérations physiologiques sur l'éclairage et applications à l'examen ophtalmoscopique", *Archives générales de médecine*, **18** (1861), 26–37.

Chapter 3
Spectral Analysis and the Telluric Lines

> *"L'Étude de la Lumière nous révèlera la constitution physique du système du monde."*
>
> *[The study of Light will show us the physical organization of the system of the world.]*
>
> Jules Janssen [1]

The house in rue Labat had a belvedere from which it was possible to make astronomical observations, which, in that particular part of Paris, was not particularly extraordinary. So this should not be taken as an outward sign that the Janssens were well-off, when in fact they were only tenants. Janssen certainly already had a lot of contacts and the outcome of the quarrel with Cima redounded to his credit; but he had no position and the couple were still living very frugally, even though there was enough money to employ a maid. It was, however, with full knowledge of these facts that Janssen chose, in 1857, to devote his life to scientific research. He knew that wealth would always be unattainable and that comfort would come only after long, hard effort, but perseverance was one of his qualities.

He had already used spectroscopy extensively in his thesis, he had read, thought, and studied a lot, and he saw a promising avenue that he could explore, at home, without great cost:

> The starting point for my research was the Earth's atmosphere. In 1833, the distinguished and venerable M. Brewster had discovered alterations in the solar spectrum that indicated that our atmosphere exerted an absorbing effect on light. These alterations consisted of the appearance of dark bands in the solar spectrum, which increased when the body rose and set. But because the phenomenon disappeared when the body was at a certain height, it was not possible to consider that their production by the atmosphere had been demonstrably proven. A direct experiment, in which MM. Brewster and Gladstone tried to reproduce the absorption phenomenon directly, had given a negative result. I followed-up this work. [2]

The Germans Kirchhoff and Bunsen had recently shown that spectral analysis allowed one to determine the chemical composition of stars [3], and Kirchhoff [4]

F. Launay, *The Astronomer Jules Janssen: A Globetrotter of Celestial Physics*, Astrophysics and Space Science Library 380, DOI 10.1007/978-1-4614-0697-6_3,

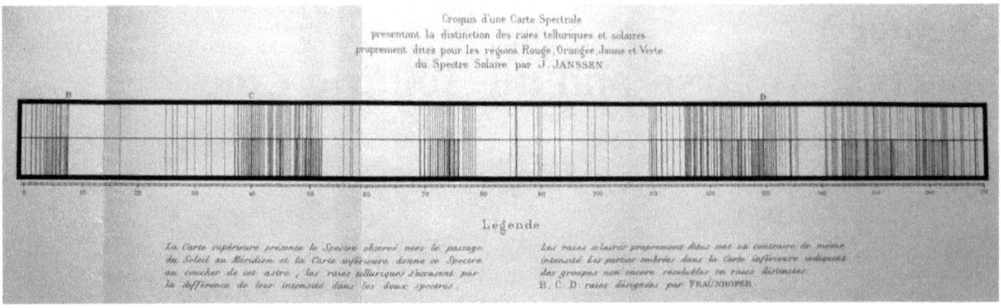

Fig. 3.1 *Telluric and solar lines by Jules Janssen*. Œuvres scientifiques. © Author's collection

had confirmed the solar origin of numerous "Fraunhofer lines"[1] [5], some of which belonged to metals. Janssen studied the spectral bands from his belvedere in Montmartre and, using a prism with a high degree of dispersion, "managed to resolve the bands observed by MM. Brewster and Gladstone into narrow lines" [6]. He observed the Sun at all elevations and determined that the intensity of the lines was continuously variable as a function of the length of their path through the atmosphere, but never disappeared from the spectrum. He deduced from this that the Earth's atmosphere was the principal cause of these bands, to which he gave the name "bandes telluriques" ["telluric bands"] [7] and concluded: "The cause of the lines in the solar spectrum is twofold: they are Solar and Terrestrial" [8] (Fig. 3.1).

So if he wanted to study the solar spectrum properly, it was necessary, on the one hand, to determine the significance of the role played by the terrestrial atmosphere by attempting to measure it as accurately as possible and, on the other, to avoid it as much as possible.

The Fine Skies of Italy and Janssen's Spectroscopes

To this end, and on the recommendation of Babinet, he again asked the Minister for support, and was entrusted with an expedition, but this time one that was financed. On 19 October 1862, he left Henriette, Antoinette and his mother for a 6-month stay in Italy. Because he needed the best possible instruments, he took the spectroscopes [9] that he had designed and which he had constructed by Jean Georges Hofmann, the Paris instrument maker of 3 rue de Buci,[2] and which had been demonstrated at the Académie des sciences on 6 October by Babinet [10]. These were powerful instruments because they consisted of a series of prisms, and particularly because they were "direct vision," that is to say, arranged in a straight line, that were easily

[1] This was the name used for the dark lines (the absorption lines that appear on a continuous background) that were observed in the spectrum of the surface of the Sun, the photosphere, ever since the scientist had drawn up an initial listing in 1814, giving the locations of the most intense, denoted A to H, with I used for the violet end of the spectrum.

[2] He later moved to 29 rue Bertrand.

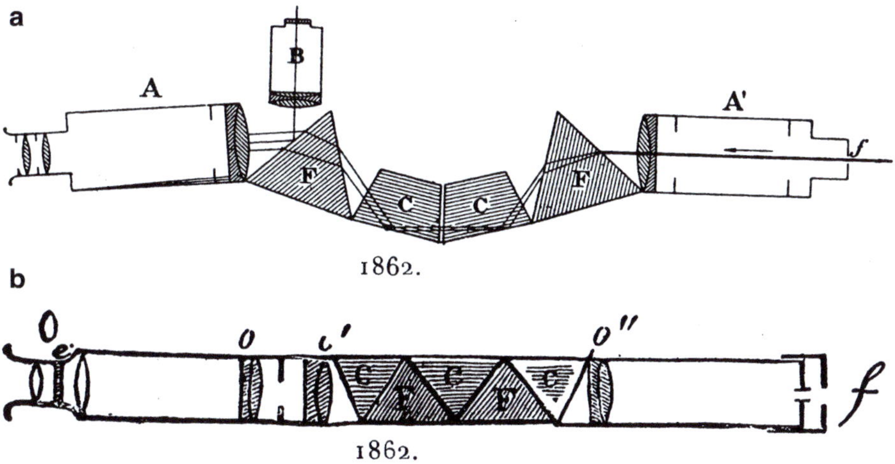

Fig. 3.2 *(**a** and **b**) Diagrams of Janssen's spectroscopes*. Œuvres scientifiques. © Author's collection

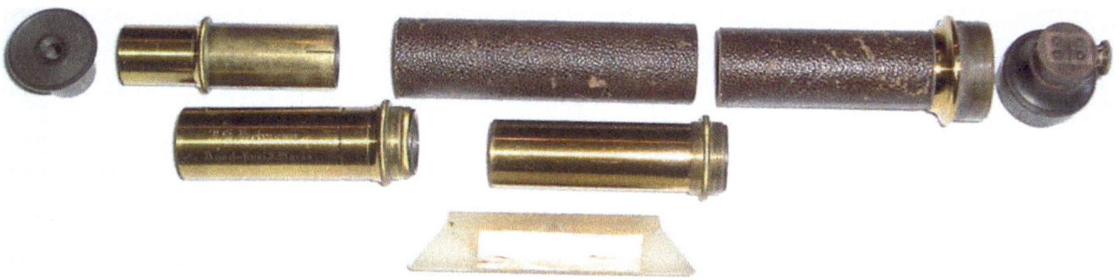

Fig. 3.3 *Janssen's direct-vision spectroscope as constructed by J.G. Hofmann*. © Author's collection

carried in a pocket. Diagrams were shown in Italy on 7 December [11] (Figs. 3.2 and 3.3).

Soon, an argument arose between Janssen and Hofmann because the latter wanted to claim responsibility for both the construction and the design of the instruments. Hofmann wrote, for example, in abbé Moigno's bulletin, *Les Mondes*:

> The spectroscope that M. Janssen normally uses is his in the sense that he paid for it. I built it by trial and error, guided by some suggestions that I owe to M. Babinet.

Janssen, who as we have seen was rather authoritative, would have had difficulty in admitting that his orders had not been carried out perfectly and, in particular, that anyone should be averse to making the necessary modifications after trials of the prototype built to his design. To cut things short, he immediately put out a statement in the same publication:

> M. Hoffman *[sic]* has constructed this small piece of equipment to my design and in accordance with my daily instructions, instructions that have been very numerous, because M. Hoffman did not understand them as clearly as might have been hoped. [...] To summarize, when M. Hoffman states that he is the inventor of this spectroscope, he abuses the confidence that I had in him, as an instrument maker, by providing him with the designs and the models [12].

Very prudently, and luckily, Janssen never forgot to describe his projects in a short note to the Académie des sciences or to a learned society. It was this that allowed him come out, with his head held high, of the dispute that arose with his scientific host in Rome, the famous Jesuit, Angelo Secchi, the astronomer at the Roman College.[3] Secchi was a recognized expert in the field – a fully justified reputation. Indeed, in 1860, at a time when there was still doubt both as to the actual existence of the luminous formations – "prominences" – observed during total solar eclipses, and also whether they belonged to the Sun (and not to the Moon), Secchi was in Spain to observe the eclipse of July 18. Four hundred kilometers away there was another observer, the very skilled amateur astronomer, printer, and chemist, Warren De La Rue, a native of Guernsey, who had studied, part of the time, in France. The two observers took excellent photographs of the event, and it was thanks to the comparison of the two series of plates that the question was answered definitively, opening the way to later studies of the Sun. It is worth emphasizing that this was certainly the first time that serious scientific conclusions could be drawn from astronomical photographs. It was, therefore, a great honor for Janssen to be invited to use his spectroscopes on Secchi's refractor at Rome.

Janssen arrived in the Italian capital in mid-November, after having confirmed, when passing through Milan and Genoa, that the telluric lines were still present under the clear Italian skies. At Florence, the weather was poor, which left him time to linger in the museums, but which was not an unpleasant occupation for an informed art-lover like him. At Rome, everything looked fine on the scientific front: "The Reverend Father Secchi, the Roman College astronomer – the leading man in Italy – received me as a colleague. He has put his large refractor, one of the largest in Europe[4] at my disposal, and I intend to use it to make some preliminary studies of stellar spectra" [13]. In practical terms, Janssen got remarkably well organized. Thanks to the official nature of his mission and the fact that "the Villa-Medici had been indicated *[to him]* as one of the most notable high points" [14], he immediately visited the French ambassador, who introduced him to the Director of the Académie de France in Rome. The next day, Janssen settled into the apartment that the latter had allocated to him, and which was no other than the one occupied by the painter Horace Vernet when he was the director of School of Rome! And it was thanks to this dream lodging that Janssen became acquainted with two holders of the Prix de Rome who were staying by rights in the Villa Medici, and who subsequently became two of his greatest friends: the sculptor Joseph Tournois and, above all, the painter Jean-Jacques Henner. The latter would become, after his wife, Janssen's most prolific correspondent.

Janssen's observations went extremely well. On 6 December, he wrote enthusiastically to Henriette:

> I've just returned from the Observatory of the Roman College, where, with my small spectroscope, I have obtained one the finest and least expected results – Spectra of the stars! – and

[3] This is the *Collegio Romano*, which was founded in 1551 by Ignazio di Loyola, and which was to become the *Pontifica Università Gregoriana* (Pontifical Gregorian University) in 1873.

[4] This was an equatorial instrument, built by Merz in 1854. The diameter of its objective was 24 cm, with a focal length of 4.3 m.

> beautiful spectra with colors and magnificent lines. Just one more step and the chemical composition of the universe will be revealed. – Here's how it happened. The Reverend Secchi, director of the Rome Observatory (and the foremost astronomer in Italy) was so surprised and pleased by the lunar spectrum that I showed him, that he suggested fixing that spectroscope to his large refractor for the stars. He was kind enough to meet the costs of the arrangement needed for it to be adapted. Then we turned the great refractor, fitted with the spectroscope, onto the star Wéga *[sic]* (in Lyra), and we saw a spectrum, but although it already exhibited colors (which had never been seen before through a spectroscope with stars), it still lacked light. Then I suddenly had an idea for a change that would give far more light and then, to our great joy, we saw magnificent spectra with indisputable lines, and which will give us marvelous knowledge of the nature of these distant Suns. The good Brother was even more enthusiastic than me. [15]

A few days later he even waxes lyrical:

> It is a wonderful sight to see astronomers in their observatories, surrounded by the silence of the night and beneath the vault of stars, using machinery to move these gigantic refractors, alongside which man appears no more than a fly. There is something mysterious about this silence, about this giant piece of dimly lit equipment, something mysterious and grandiose, which lifts the spirit and carries it enthusiastically on a quest to probe the very essence of the great laws of the Universe! ... [16]

During the Christmas holidays, Janssen happened to catch sight of the pope, to whom he had refused to be presented officially, because Pius XI supported a government of which Janssen did not approve: "It is unfortunate that a man like him, who is very good and very well-intentioned should serve such a bad cause, and should, all in all, do so much harm to humanity's greater interests" [17], he confided to Henriette.

Janssen did a lot of work on stars, and on 24 January 1863, he determined the presence of atmospheric lines in Sirius, which confirmed his theory. In addition, he discovered the presence of sodium in the principal star of Orion:

> I was able, repeatedly, to establish complete agreement between the D line in a flame and the dark line in the star [...] and it seemed appropriate to call attention to a fact that is, undoubtedly, our first step in the study of the chemical composition of our nebula. The accuracy of this conclusion has since been confirmed by MM. Miller and Huggins (1865). At the time when I made this observation it was one of the first facts *[establishing]* the unity of the material elements in the system of the world. [18]

The stay in Italy looked likely to continue, and Henriette obviously found that time was dragging: she complained about not having been able to accompany her husband, whose vocation as a disinterested researcher became more and more evident in the letters that he sent to her:

> We must, above all, blame our lack of fortune, in particular, for the fact that we could not travel together. But when I see the irritating effects of a considerable fortune, I may perhaps thank the Lord that he has seen fit to ordain that I should follow a harder and narrower path, but one that is more fertile and honorable. Man should experience toil and sweat; these are his sanctification and his salvation. I have not been spared this task; I pray to God that it should become more and more a need for my existence, and if it pleases the Creator to guide this activity towards wholesome matters, of extreme utility, all my wishes will be granted. [19]

At the beginning of March 1863, Janssen decided to leave for Naples where he hoped to find a refractor. In fact, Secchi's was no longer available to him, and

Janssen understood the real reason for this a fortnight later. Father Secchi had, in effect, betrayed him:

> I had, my good Henriette, the great idea of publishing the results I have achieved as soon as possible at the Academy in Rome – Everything had been printed at the Academy's expense, and I had been given 50 copies – There was part of the work on stars that could not be *[published]*, because it was not completely finished. The Reverend Father Secchi, jealous of me for obtaining so many results, and having, moreover, learned from me how these studies were carried out, suddenly informed me that he could no longer lend me his refractor. He set to work to finish *[the observations]* himself, and published them in a German journal, not daring to do so at Rome or Paris. This disloyal conduct has aroused indignation here, because my friends here have let it be known. I, however, have not felt it appropriate to promote a public scandal. I have, in Paris, the means of proving his treachery and exposing him and, to reassure you, I will tell you that the major portion is already mine through the publications that have already appeared. The Reverend Father is a priest and Jesuit.
>
> He's a specimen of the majority of the Roman clergy. Ah! My dear child, if France had had to undergo what has to be supported here, it would not have lasted long. The Government knows that, but the general populace does not! You can see, my dear friend, that I am doing good work, because men in higher positions do all that they can to take the credit for themselves. That should give you confidence in your friend – envy and jealousy are, in fact, the highest compliments! The Secretary of the Academy, who detests Father Secchi, will tell the Academy about the matter, in which he will recount how it was I who introduced Rome to celestial spectral analysis, and that the methods of observation are thanks to me. [20]

Janssen returned to Paris in April, but the matter of the telluric lines was far from being resolved.

The Faulhorn and Lake Geneva

Whereas he had a credit of 1,200 F to go to Italy, in June 1864 [21] he was allocated 2,000 F by the Emperor Napoleon III's new minister for public education, the historian Victor Duruy, to travel to the Bernese Alps. Initially, he went to the summit of the Faulhorn to continue his investigations:

> Ascending a high mountain, which allows one to leave a significant portion of the atmosphere below one, should result in diminishing even more the absorption phenomenon that concerns us: and this is what I have observed. During the week that I stayed at the summit of the Faulhorn, at an altitude of nearly 3,000 m, I noticed a general decrease in the solar spectrum of all the dark rays of terrestrial origin. [22]

The penultimate, decisive experiment to be carried out was reproducing the telluric lines with artificial light sources. Janssen went to the shores of Lake Geneva, profiting from the great depth of water vapor that permanently lay above it.

> The flames of a large bonfire of pine-wood on the pier at Nyon was studied from the tower of the church of Saint-Pierre at Geneva. Close by, the flames did not exhibit any specific alteration; the spectrum was perfectly well-known and continuous, whereas at Geneva, 21 kilometres from the bonfire at Nyon, the spectrum displayed the lines observed by Mr Brewster at sunset. [...] I then asked myself to which elements in our atmosphere we

> should attribute this remarkable phenomenon. During the course of these studies, I was led to [...] attribute this phenomenon to the water vapor that is widespread in our atmosphere. [...] But there still remained one direct experiment to be carried out to demonstrate the weight of these predictions. [23]

In the meantime, Janssen submitted a report to the Académie des sciences and entered the Bordin Prize, the subject of which that year should relate to the "theory of optical phenomena." The memorandum "which particularly caught the attention of the Commission" [24] – which consisted of MM. Pouillet, Foucault, Edmond Becquerel, Babinet, and Fizeau (the rapporteur) – was identified as number 6, with its topic given as "The study of Light will show us the physical organization of the system of the world." After having given a brief description of the six memoranda that had been submitted anonymously to the Commission, the rapporteur stated, regarding the session devoted to that particular submission that dealt with "the telluric lines in the solar spectrum":

> Your commissioners were unanimous in recognizing the distinctive merit of this work, and in considering it superior to those that we had previously analyzed; however, the majority were of the opinion that several significant points needed to be covered to show the degree of rigor and the extent of the developments that are appropriate to the subject. [25]

As a result, the sum of 3,000 F was divided between three candidates, and "an award of 1,500 F is made to the author of memorandum number 6." In other words: It was good, but the applicant could and should do better...

The Gasworks at La Villette

Janssen had no laboratory. But never mind: he had contacts! In November 1865, he introduced a course in general physics at the École spéciale d'architecture, a post that he retained until 1871, and it is worth spending a moment on the opening speech that he gave. It was, in fact, the chance for him to express, loudly and strongly, the place that physics should take alongside artistic subjects:

> Let us never forget, however, that the architect is an artist and should remain so. This is where the difficulties lie with your syllabus. Science and art, isn't that a lot to ask? Also has one not sometimes wondered whether such varied knowledge can really be held in a way that will allow it to be applied fruitfully, and if there might not be an advantage, leaving the field of art to the architect, if the engineer were not to be given everything in construction relating to science. A moment's thought is enough to show that such a division is impossible. Indeed, how could an architect, entirely deprived of any scientific knowledge, take into account, within the overall scheme of his work, the requirements that only science can answer nowadays? [26]

And Janssen went on to discuss at length the aims of the programme by describing certain subjects where physics was involved: heat and the problems of expansion and thermal conductivity of materials; electricity and lightning conductors; acoustics and sound insulation for auditoria and meeting halls, as well as urban dwellings; and finally, light and the problems of industrial and commercial lighting, in particular that for locations devoted to sales.

It was the Director of the École who put Janssen in touch with the director of Compagnie parisienne des gaz *[Parisian Gas Company]* and, in 1866, the gas works at la Villette became the location for his experiments:

> A sheet-metal tube, 37 metres long, buried in a box full of sawdust, and closed at each end by strong glass windows, was filled with water vapor at a pressure of 7 atmospheres, from a boiler at the works. [...] A beam of light provided by a set of 16 gas jets passed through the tube, and could be analyzed on exit. The vapor imposed on the light most of the changes that had been observed with the terrestrial atmosphere. [27]

The conclusion to be drawn was clear: it was water vapor, rather than any other element in the air that acted to produce the telluric lines because "the effects of the length of the tube and those of pressure were investigated. The lines developed as the length increased and as the pressure was raised; they decreased when the opposite occurred" [28]. And now that the spectrum of water vapor had been analyzed in the laboratory, it would subsequently be possible to express with certainty, rather than theories, about its presence in the universe:

> It is, above all, in astronomy where it will be interesting to develop the results. Based on accurate knowledge of the spectrum of water vapor, I am confident that I will soon be able to give an opinion about the presence of this essential element for organic life in the atmospheres of the planets and other celestial bodies. Even today, I can state that water vapor is not part of the atmosphere of the Sun. [29]

During the meeting of the British Association for the Advancement of Science (BAAS) at Nottingham in August 1866, "M. Huggins made the public announcement of the discovery that had just been made on water vapor, 'a discovery made by a distinguished visitor present at this meeting'" [30] Janssen proudly reported to his wife. William Huggins (Fig. 3.4) was a recognized authority in astronomical spectroscopy, and Janssen would thenceforth give great attention to his publications, which Henriette translated for him.

He was happy, and forgave all his family for everything when he ended his letters:

> Lots of love:
>
> 1. Mother (a son's heart full of respect)
> 2. Mme Janssen (a spouse's heart that recalls only the good)
> 3. Mlle Antoinette (a father's heart, blind to faults, and a Lynx when it comes to good qualities). [31]

At Nottingham, he stayed with an English engineer, whose daughter took him on a trip in a carriage, with a professor from Edinburgh to the home [*Burgage Manor*] of Lord Byron: "There are constant invitations and counter-invitations. And with all that, serious matters still go ahead" [32]. Indeed, it was suggested to Janssen that he should be nominated as a foreign member of the Association, dispensing with subscriptions. In London, he stopped to see Wheatstone, the physicist, who had invited him; the Vice-President of the Royal Society, Gassiot; and also to visit the solar observatory at Kew, where he very probably saw De La Rue.

Fig. 3.4 *Sir William Huggins*. Astrophysical Journal, 1907. © Observatoire de Paris Library

The Trani Eclipse

An indefatigable traveler, Janssen left again in February 1867 on an expedition to observe an annular eclipse the following month at Trani, in Italy, where crowds besieged him:

> At the moment of the eclipse the house where I was, was invaded and it was necessary to send for the guard to prevent the rush. They took over my instruments, they jostled me, insisted on looking through the large refractor: 'Mossiou *[Monsieur]* let us see; Mossiou, when will the Moon be in front of the Sun; Mossiou, what are you looking at there; Mossiou, Mossiou! [...] But none of these hordes suspected that it was serious research, they believed that they had the right to see through my instruments. [33]

The serious research concerned comparing the spectrum of the center of the Sun with that at its limbs. By placing the slit of the spectroscope, at the instant of totality, at the point where solar lines had been confirmed, Janssen was hoping to detect an increase in the strength of the dark absorption lines at the limb, which would prove the existence of a significant solar atmosphere, at a high temperature, "surrounding a solid or liquid core, at the highest temperature one could imagine" [34], and thus even higher than that of the atmosphere, as Kirchhoff expressed it in "the most probable hypothesis." Consequently, he used a five-prism spectroscope to follow several faint lines of iron in the Fraunhofer spectrum. He was lucky because the weather cleared fortuitously at the time of the eclipse, but "these lines did not show any detectable increase" [35]. Janssen, who remained cautious [36] because he had not been able to observe the Sun's extreme limb, was in fact very satisfied with these results, which tended to invalidate Kirchhoff's theory and confirm that of the "Master," Hervé Faye, professor at the École polytechnique, which ascribed the source of the non-telluric Fraunhofer lines to the photosphere itself. Faye did indeed consider "the Sun as an essentially gaseous globe" with its own extremely high temperature, whose "radiation to celestial space has caused cooling of the surface, which, through condensation, has altered the gaseous elements in these regions into the state of a solid dust or a liquid. [...] So, through a relative reduction in the temperature, the gaseous globe is surrounded by a very luminous envelope: this is the photosphere,[5] the visible portion of the Sun" [37].

The Study of Volcanoes and the Invention of the Brush Thermometer

From Trani, Janssen traveled to Greece, and in April 1867 he was on the island of Santorini to study, by spectral analysis, the composition of the gases emitted by the "flames of its volcano" [38]. The experiments were far from being easy to carry out, and others were even more risky: to make magnetic observations, Janssen did not hesitate to put himself at risk from "blocks of lava ejected higgledy-piggledy, with burning and mephitic vapors and a wind that was frequently irresistible" [39].

It was for this expedition that Janssen invented the "brush thermometer." This clever instrument, which was designed to obtain the temperature of the surface of sea and river water, was considered, in 1872, by its inventor to be "a gain for science."

> The new arrangement with this instrument consists of the fact that the bulb is held in the centre of a brush, consisting of threads of hemp. This brush [...] which is fixed to the framework of the thermometer, carries a lead collar. When the instrument is thrown into the water, the lead collar pulls it down rapidly and vertically; the hemp threads immediately part, and the thermometric reservoir is then in contact with the liquid, and takes its temperature. Equilibrium is reached in a few seconds, and the thermometer may be withdrawn. [...] As soon as the instrument is out of the water, the threads close on one another, surrounding the bulb and, by capillary action, retain a considerable quantity of the liquid of which one requires the temperature. The presence of this liquid [...] enables one to read the scale at one's leisure. [40]

Warming by ambient air is compensated by cooling caused by evaporation, and Janssen ascertained that the time required for reading was at least a quarter of that required for warming to begin ...

[5] This name is quite old, because it was used by the German, Fischer, in May 1788: Fischer, "Ueber die Sonnenflecken" in *Astronomisches Jahrbuch für das Jahr 1791* (Berlin, 1788), 195–201, p. 197.

Water Vapor in the Martian Atmosphere and Life in the Universe

Janssen then left for Sicily, and after having made observations on Etna, staying at the summit for 3 days, at Palermo and then Marseille in the return trip, he was in a position to announce the presence of water vapor in the atmospheres of Mars and Saturn! And so, "that there are powerful reasons for thinking that life is not the exclusive privilege of our small earth, the younger sister of the great family of planets" [41].

When, in 1896, Flammarion asked Janssen for clarification on some historical points, the reply was not long in coming, and Angelo Secchi's perfidy resurfaced:

> As far as the "minor, sensitive point" is concerned, believe me that it is nothing of the sort, and that this point of scientific history is perfectly simple and clear. I discovered the telluric lines in 1862 (*CR* June 1862) but did not announce the presence of water vapor in Mars until 1867 (*CR T.* **64**, p. 1304) for the very good reason that until 1866 (*CR T.* **63**, p. 289) I did not know the spectrum of water vapor, the spectrum being given by the experiment using the tube, 37 metres long, at a pressure of 6 atmospheres, at the works at La Villette. One would not scientifically know *[enough]* to announce the presence of water vapor in any planet before being in possession of this spectrum. Just as one could not announce the presence of iron in the solar atmosphere without knowing the spectrum of iron. But this is just what Father Secchi did. [42]

Research Trips or Tourist Travels?

Having returned in June 1867, Janssen left again in July for the Azores, where he was sent by the Académie des sciences on an expedition with his great friend, the geophysicist Charles Sainte-Claire Deville (Fig. 3.5),[6] to again dedicate his time to "applying physical research to the study of an active volcanic vent" [43]. Unfortunately, the eruption that had begun in June had ceased. Janssen contented himself with magnetic studies, but not forgetting a small detective investigation of exact witness accounts of the events of the preceding months.

Janssen then extended his trip to Portugal, where Deville left him, and then to Spain. Henriette obviously found the time very, very long, and when her husband wrote to her at the end of September to tell her about his tourist trip to Grenada, she exploded:

> It will soon be six months that I have been living with an old woman and a child, fighting against the high cost of living to try to make ends meet, having no entertainment save what I can give a six-year-old child, strictly observing the various prohibitions that you have placed upon me. It was exactly the same during your trip to Italy, to Brittany, and England, and I have had the grief, despite that, of seeing that you have almost always deliberately doubled the length of your trips. […] As far as travel is concerned, you make me live in

[6] The brother of the chemist Henri Sainte-Claire Deville.

Fig. 3.5 *Charles Sainte-Claire Deville. Drawing by Jules Janssen.* © Library of the Institut de France

> hope, until I am so old as to be unable to do so. [...] It seems to me that you could see the [whole of] south of Spain in a fortnight. Believe me, Jules, you neglect far too much *[doing anything]* that make your wife happy, and to satisfy her tastes to the best of your ability; this is not the way to keep her young, happy, and in good health, and it seems to me that should be part of your own happiness. There are men who leave their wives for mistresses; you do it for journeys! [44]

Farther on, she again mentions her problems with money: "You only left me just enough money for 3 months. The needs of the winter are beginning to be felt" and she recounts a revealing "saying" of little Antoinette's "I'm as cold as a poor man"! The soup that the child was bored with obviously had not been enough to warm her.

Janssen would probably have been more annoyed than affected by these recriminations from his wife, but he was certainly extremely proud of the congratulations that the Minister, Victor Duruy, sent him on the following 22 December, on receipt of his report:

> I have just read your memorandum with the greatest interest. You are following a fertile path. Your research will eventually bring major results. Accept my congratulations and believe me willing to help you in this task [45].

References

1. Janssen, Jules, "Mémoire sur les raies telluriques du spectre solaire (juillet 1865)", in *Œuvres scientifiques*, vol. I, 81–96, p. 81.
2. Janssen, Jules, "Rapport sur une mission en Italie, dans les Alpes et en Grèce (1867)", in *Œuvres scientifiques*, vol. I, 143–68, p. 148.
3. Kirchhoff, Gustav and Bunsen, Robert, "Analyse chimique fondée sur les observations du spectre", *Annales de Chimie et de Physique*, **62** (1861), 452–86 and **64** (1862), 257–311. These are translations of articles published in Germany two years earlier.

4. Kirchhoff, Gustav, "Recherches sur le spectre solaire et sur les spectres des corps simples", *Annales de Chimie et de Physique*, **68** (1863), 5–45.
5. Fraunhofer, Joseph, "Bestimmung des Brechungs - und Farbenzerstreuungs - Vermögens verschiedener Glasarten in Bezug auf die Vervollkommnung achromatischer Fernröhre", *Denkschriften der königlichen Akademie der Wissenschaften zu München*, **5** (1817), 193–226; Pasachoff, Jay M., and Suer, Terry-Ann, "The Origin and Diffusion of the H and K Notation," *Journal of Astronomical History and Heritage*, **13–2** (2010), 121–127.
6. Janssen, Jules, "Rapport sur une mission en Italie, dans les Alpes et en Grèce 1867", in *Œuvres scientifiques*, vol. I, 143–68, p. 148.
7. Janssen, Jules, "Note sur les raies telluriques du spectre solaire", *CRAS*, **54** (23 June 1862), 1280–1.
8. Janssen, Jules, "Mémoire sur les raies telluriques du spectre solaire (juillet 1865, in *Œuvres scientifiques*, vol. I, 81–96, p. 96.
9. Aubin, David, "Orchestrating observatory, laboratory, and field: Jules Janssen, the spectroscope and travel", *Nuncius*, **17** (2002), 615–33.
10. Janssen, Jules, "Notes sur trois spectroscopes", *CRAS*, **55** (6 October 1862), 576–8.
11. Janssen, Jules, "Notes sur de nouveaux spectroscopes par M. J. Janssen de Paris", *Atti della Pontifica Romana dei Nuovi Lincei*, **16** (1862), 73–5.
12. Janssen, Jules, "Réponse de M. Janssen à M. Hoffman", *Les Mondes*, 1 (16 July 1863), 636.
13. Letter from Jules to Henriette Janssen, 19 November 1862, BIF Ms 4133–43.
14. Letter from Jules to Henriette Janssen, 13 November 1862, BIF Ms 4133–42.
15. Letter from Jules to Henriette Janssen, 6 December 1862, BIF Ms 4133–45.
16. Letter from Jules to Henriette Janssen, 12 December 1862, BIF Ms 4133–46.
17. Letter from Jules to Henriette Janssen, 31 January 1863, BIF Ms 4133–52.
18. Janssen, Jules, "Rapport sur une mission en Italie, dans les Alpes et en Grèce (1867)", in *Œuvres scientifiques*, vol. I, 143–68, p. 146–8.
19. Letter from Jules to Henriette Janssen, 24 February 1863, BIF Ms 4133–54.
20. Letter from Jules to Henriette Janssen, 18 March 1863, BIF Ms 4133–58.
21. Ministerial order of 24 June 1864, AN F^{17} 2977.
22. Janssen, Jules, "Rapport sur une mission en Italie, dans les Alpes et en Grèce (1867)", in *Œuvres scientifiques*, vol. I, 143–68, p. 149.
23. *Ibid.*, pp. 150–1.
24. "Prix Bordin, rapport sur le concours de 1865", *CRAS*, **62** (05 March 1866), 485–90.
25. *Ibid.*
26. Janssen, Jules, "Discours prononcé à l'ouverture du cours de physique générale à l'École spéciale d'architecture at Paris (November 1865)", in *Œuvres scientifiques*, vol. I, 97–113, p.99.
27. Janssen, Jules, "Rapport sur une mission en Italie, dans les Alpes et en Grèce (1867)", in *Œuvres scientifiques*, vol. I, 143–68, p. 151.
28. Janssen, Jules, "Sur le spectre atmosphérique terrestre et celui de la vapeur d'eau" in *Report of the 36th Meeting of the BAAS, Nottingham*, August 1866, 11.
29. Janssen, Jules, "Sur le spectre de la vapeur d'eau", *CRAS*, **63** (13 August 1866), 289–94.
30. Letter from Jules to Henriette Janssen, 26 August 1866, BIF Ms 4133–74.
31. Letter from Jules to Henriette Janssen, 22 August 1866, BIF Ms 4133–73.
32. Letter from Jules to Henriette Janssen, 27 August 1866, BIF Ms 4133–75.
33. Letter from Jules to Henriette Janssen, 06 March 1867, BIF Ms 4133–79.
34. Kirchhoff, Gustav, "Recherches sur le spectre solaire et sur les spectres des corps simples", *Annales de Chimie et de Physique*, 68 (1863), 5–45, p.30.
35. Janssen, Jules, "Sur la constitution du Soleil", *CRAS*, **73** (14 August 1871), 432–6.
36. Janssen, Jules, "Observation de l'éclipse annulaire de Soleil du 6 mars à Trani (Italie)", *CRAS*, **64** (18 March 1867), 596–7.
37. Janssen, Jules, "Résumé des notions acquises sur la constitution du Soleil", *CRAS*, 68 (08 February 1869), 312–4.
38. Janssen, Jules, "Sur la composition des gaz émis par le volcan de Santorin,", *CRAS*, **64** (24 June 1867), 1303.

39. Letter from Jules to Henriette Janssen, 20 April 1867, BIF Ms 4133–82.
40. Janssen, Jules, "Nouveau thermomètre destiné à prendre les tempèratures de la surface des eaux marines ou fluviales", in *Report of the 42nd Meeting of the BAAS, Brighton, 1872*, 59–61.
41. Janssen, Jules, "Rapport sur une mission en Italie, dans les Alpes et en Grèce (1867)", in *Œuvres scientifiques*, vol. I, 143–68, p. 152.
42. Letter from Jules Janssen to Camille Flammarion, 28 November 1896, SAF, Fonds Flammarion de Juvisy.
43. Janssen, Jules, "Sur un voyage fait aux Açores et dans la péninsule ibérique", *CRAS*, **65** (14 October 1867) 646–7.
44. Letter from Henriette to Jules Janssen, 27 September 1867, BIF Ms 4134–57.
45. Letter from Victor Duruy to Jules Janssen, 22 December 1867, BIF Ms 4135–94.

Chapter 4
Janssen and the Solar Flames: The Key Eclipse of 1868

"On m'a envoyé pour observer l'éclipse pendant 5 minutes et je leur rapporte des Grandes Indes l'éclipse perpétuelle."

[I was sent to observe the eclipse for 5 minutes and I am bringing back for them from the Indies an eternal eclipse.]

Jules Janssen [1]

The request for help did not take long to arrive at Duruy's office because the longest total solar eclipse was due to occur on 18 August 1868. Whereas an ordinary eclipse lasts between 2 and 3 min, the duration of this one would actually be 6 min 57 s because the Sun would be almost as far from the Earth as it can be, and the Moon almost as close as it can approach. The conditions were thus exceptionally favorable for achieving spectral analysis of the Sun's "flames," the prominences, of which Janssen wrote "the colour is sometimes red, sometimes pink, violet, peach, white, and even black" [2]. In fact, if "the prominences consist of solid material, they will give a continuous spectrum. Are they incandescent gas? The spectrum will consist of brilliant lines, and the location of the lines will reveal, to an experienced eye, the nature of the gas" [3]. This is what spectral analysis would enable one to determine. But Janssen had only just enough time to prepare, because the line of visibility of the eclipse's totality lay across Asia, and it would be necessary to leave 2 months before the event.

Preparations

While the British were getting organized to observe from India, the Germans from Aden, and the Dutch from Celebes, France, which had initially thought to entrust observations to seamen, rather than astronomers [4], vacillated. Being independent of any institutions and, in particular, avoiding the authoritarian influence of the director of the "Imperial" Paris Observatory, Le Verrier, Janssen dealt directly with

F. Launay, *The Astronomer Jules Janssen: A Globetrotter of Celestial Physics*, Astrophysics and Space Science Library 380, DOI 10.1007/978-1-4614-0697-6_4,

Fig. 4.1 *Victor Duruy*. Album Félix Potin. © Author's collection

the minister and his department that was responsible for expeditions. The order that concerned him, and which sent him to British India, with a grant of 15,000 F, was signed by Duruy on 9 March [5] (Fig. 4.1).

So, when Le Verrier approached him, Janssen could take pleasure in telling him that it was too late:

> I have received the letter in which you do me the honour of asking me to take part in a committee relating to the Eclipse next August. I myself, Monsieur, must leave immediately for India, most particularly for the plateaus of High Asia, where I intend to carry out research into both celestial and terrestrial Physics before observation of the eclipse. Being entirely involved in preparations for this major voyage, it would be extremely difficult for me to carry out the work of such a commission. I deeply regret not being able to give yet again this proof of my devotion to our French Science, but I hope that it will be considered that I am paying my debt in another form. [6]

A few days later, Duruy shook Le Verrier:

> … let us try, by actual activity, to make up for these really annoying delays. The location chosen will save time. Can you not find the means of adding spectroscopy and photography to the purely astronomical observations [7],

and he matches this sensible piece of advice with another that is no less significant: "A piece of good advice – Speak less ill of me [*to the emperor*], that does no good." Indeed, Napoleon III, who had followed Duruy's university career with close attention, had entire confidence in his minister: "No minister ever had so much liberty and was so little shackled by interference from the sovereign in his internal administration" [8] Duruy himself was to say.

Finally, Le Verrier, who had suffered plenty of other refusals, sent Édouard Stephan, the young director of the Marseille Observatory, accompanied by his colleagues Georges Rayet and Félix Tisserand, to Siam (now Thailand), on the Malacca peninsula, where they were met and accompanied by the king himself [9]. The decisions were not taken without bitter tensions. Duruy had to intervene once more, and signed the order for the expedition on 29 May, that is, 2 months after Janssen's. The French team for Siam had the remarkable credit of 50,000 F, the third of the Paris Observatory's annual budget.

Because the date of the eclipse fell in the middle of the monsoon, "stations that offered the most favorable chance were those that proved to be sheltered by a high chain of mountains and the width of a major continent" [10]. Janssen decided quickly: "As soon as the Bureau des longitudes and the Academy had done me the honour of selecting me, I had no hesitation in choosing the eastern coast of Hindustan. The station should be Masulipatam or nearby," not far from Madras.

Leaving Paris on 16 June, Janssen took ship from Marseille on the 19th, with the members of the other French expedition, from whom he parted in Ceylon. On his arrival in Madras, he was received with the greatest courtesy by the British authorities, partly thanks to the letters of recommendation from Warren De La Rue, with whom he maintained extremely friendly relations. Despite the difficulties arising from the moist heat and from the considerable amount of material that had been imported (twice what was necessary, as a precautionary measure), Janssen arrived 2 weeks early at Guntoor, the observing location finally selected, equidistant from the mountains and the sea. There he found a family of French merchants, who had long been established in the Indies, the Lefaucheurs, who offered him the whole of the first floor of their house, the highest and best sited in the place, which was linked to a terrace. It was thus there that Janssen observed the eclipse. "I am having a screen of bamboos and mats made against the wind. We have the whole of an immense room for our instruments. These families are proud and happy to receive us" [11], he wrote to Henriette on 4 August.

Observation of the Eclipse

Janssen had four large 16-cm refractors at his disposal, together with a 21-cm Foucault reflector, the mirror of which had been parabolized, free of charge, by Martin, the optician. The refractors were cleverly interlinked, and the overall motion was provided by a mechanism constructed by the brothers Brunner. Janssen, whose only assistant was a cadet from the liner *l'Impératrice*, made the most of the abilities of three members of the Lefaucheur family, whom he introduced to the tasks of drawing and taking measurements!

It was known at that time that the lines that Fraunhofer called C (in the red) and F (in the blue) were part of the spectral signature of hydrogen, its "identity card," while the D line (in the yellow) corresponded to sodium, and was in fact, two close lines D_1 and D_2, which had not been separated [*by Fraunhofer*] in 1814.

All the observers whose spectroscopes were directed at the assumed position of the prominences, observed bright lines, of greater or lesser intensity, and characteristic of gaseous emission. Whereas among the Britons, Lieutenant John Herschel (son of Sir John and grandson of Sir William) saw three only (one of which was in the red, closer to B than C; an orange one exactly on D; and a blue line almost on F); Norman Pogson saw five (one of which was yellow, on or near D, but no red); and Colonel James Tennant (who was also at Guntoor) again five (one red, one yellow, and one blue corresponding exactly with C, D and F), but with considerable difficulty; the Frenchman Rayet beat all records despite the absence of the red line corresponding to C, observing nine (of which one yellow line corresponded to D). Janssen himself saw only "five or six," and moreover he only cites five colors, without specifying their exact positions:

> Two spectra, each consisting of five very brilliant lines (red, yellow, green, blue, and violet) filled the spectral field, and replaced the prismatic image of the Sun, which had just disappeared. [...] These two spectra were caused by two magnificent prominences. [...] One of them [...] resembled the flame from the fire of a forge. [...] The right-hand prominence [...] presented the appearance of a range of snow-covered mountains ... [12]

The Discovery of the "Prominence Method"

What in fact struck him was, above all, the exceptional intensity of the lines, their "brilliance" in the blue and in the red because analysis of the light "immediately showed him that it consisted of an immense column of incandescent gas, primarily consisting of hydrogen" [13]. He soon noticed that these lines were in the exact position of the C and F lines, and it was those lines alone and, in particular, in the red C line, in which he was interested. He also noted the similarities in the chemical composition of the two prominences observed during the eclipse: their spectra corresponded line for line. At the moment when the weather clouded over, just after the eclipse, he exclaimed "I shall see these lines outside eclipses" [14]: he had just discovered a method of observing prominences and the region immediately around the Sun "at any time, without it being necessary to have recourse to interposing an opaque body in front of the Sun" [15]. It was during the following night that he perfected this. Here is how Janssen, as a lecturer, explained it during the course of a meeting at the Sorbonne in 1870:

> For a long time people had tried to see prominences outside eclipses. Stops at the focus of refractors, coloured glasses, etc., have all been tried without success. The intense light from the photosphere has always swamped the weak light from the prominences, above all because of the illumination of our own atmosphere around the disk of the Sun. [...] But let us interpose a prism. [...] The light from the Sun is approximately white; it is rich in all the rays *[of the spectrum]*, it is thus spread out by the prism. The prominences, by contrast, contribute just a few simple lines,[1] their light is not weakened ... [16]

The very next day he implemented "the method" and was able to follow the evolution of prominences with time, his eyes being fixed on the C line of hydrogen, which we nowadays know as the Hα line.

---

[1] We would nowadays say "monochromatic lines."

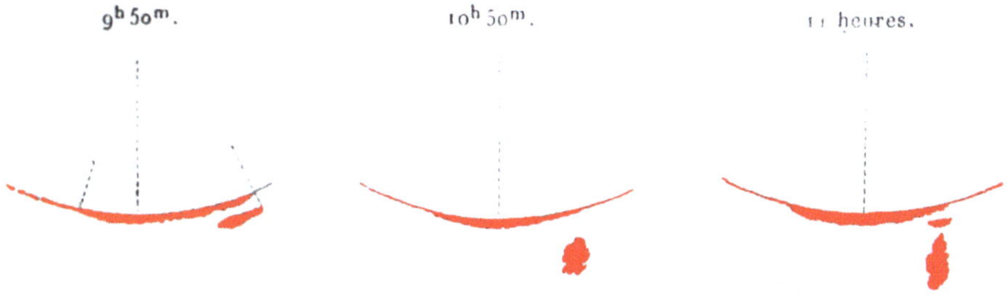

Fig. 4.2 *A prominence followed by Janssen on 4 September 1868.* Œuvres scientifiques. © Author's collection

> I was able to witness phenomena revealed by a new eclipse that lasted all day. The prominences seen the day before were greatly altered. [...] From that day until 4 September, I constantly studied the Sun by this method. I have drawn charts of the prominences [...]. This period [...] was like an eclipse that lasted seventeen days. [17] (Fig. 4.2)

Problems of Communication

Sent the day after the eclipse, the 19th, at midday, the telegram that Janssen wrote in English and sent to his wife arrived in the morning of the 21st. But because Janssen had used several French words, in particular "protubérances" instead of the English word "prominences," the Indians had problems transcribing it. The resulting text that was transmitted was "Eclipse observée protu Berankees spectrum remarkable and unexepcted [*sic*] protuberances gaseous nature" [18], and although a bit surprising (after all, *Le Moniteur* translated it as: "eclipse observed at Rotaboranas"!) it was perfectly comprehensible to scientists. It was thus known in France that it has been fine at Guntoor, and that the Janssen expedition had been a success. Henriette passed on the information to the French authorities and to British friends, including Warren De La Rue, who thus received news from Janssen before that from Tennant [19].

Everyone certainly wanted to know more, but Janssen remained astonishingly silent. As we have seen, he was observing without a break, and drawing up charts, and he only decided to write to his wife, to the Académie des sciences, to the Bureau des longitudes, and to several friends, a month later, on 19 September, from the port of Cocanada:

> I have just arrived from the interior.[2] The eclipse was observed as you will have learned from the telegram; but the most important thing, for which we should thank God, is that I have discovered a method of studying prominences outside eclipses. [...] This method will be epoch-making in science; to a certain extent it renders eclipses superfluous, and enables studies to be carried out at any time. I am writing to my dear mother, who will be extremely happy with this discovery by her son. [20]

[2] Guntoor, where he had observed the eclipse.

Arriving at Calcutta at the beginning of October, he was more concerned to talk about his hunting exploits during the trip: "On the way I shot storks, herons, and water fowl. I also killed some enormous bats, the bodies of which looked like those of foxes, and with terrible bites. I killed more than fifteen [21]," before mentioning the problems of disseminating of his discovery, although subsequent events showed that he should perhaps have given it more consideration: "If our beloved France can be foremost in the observation of that eclipse, thanks to this discovery, which it is improbable that others will have made, I shall be very happy and rewarded. Now I need to make it widely known."

In fact, thanks to the (normal) delays in the transport of post from India, his letters of 19 September did not reach Paris until 24 October. The contents were made known to the members of the Académie des sciences at their meeting on the 26th, both by the permanent secretary and by Charles Sainte-Claire Deville.

Norman Lockyer Enters the Picture

Another communication on the same subject had, however, just been made at the same meeting by Warren De La Rue. The latter informed the president of two letters that he had received, one of 21 October from the physicist Balfour Stewart, who announced that on 20 October, in London "Lockyer scored a success; he found red flames with his new spectroscope" [22], the instrument having sufficient dispersion, and which he had finally received a few days earlier. The second letter, dated 23 October, from Norman Lockyer himself, gave details. From this, it was learned that Lockyer had put forward the idea of the method 2 years earlier, on 11 October 1866, in a paper communicated to the Royal Society in London [23] ...

Very wisely, Faye, in full support of Janssen, immediately calmed the situation:

> It is certain that the initial idea of the method by which M. Janssen first, and then M. Norman Lockyer have succeeded, one in India on 19 August, and the other in England on 20 October, to capture and measure previously invisible phenomena by means of spectroscopic analysis, was devised and suggested first by M. Lockyer, but that this did not lead to any results. [...] It was only when he knew, through French and British observers[3] of the eclipse, the detailed nature of the spectrum of prominences, that he succeeded in finding signs of this spectrum in England [...]. M. Janssen, for his part, found himself face to face with the revealing phenomenon of the eclipse; he knew immediately how to interpret it; he succeeded being a true master as well, and was the first to discover what had long been sought, but fruitlessly, before him. [...] But instead of trying to proportion the merit of the discovery, and consequently diminishing it, would it not be better to impartially attribute the whole honour to both of these two men of science [...]? [24]

Charles Sainte-Claire Deville immediately went to see Henriette. He recounted to her a comment about her husband that Faye had made to him on leaving the meeting: "This man has a real genius for observation" [25], and he was anxious to be the first

[3] Lockyer had, in fact, read Rayet's report the day before.

to tell her that Duruy would be sending Janssen the cross de la Légion d'honneur within 3 days … Jules would not have had to wait for very long following the recent, unsuccessful proposal by 40 academicians, which Henriette had told him about on 15 August, and to which he had replied: "You are wrong to bother. Although according the red ribbon all the justice and respect it merits, I consider that the decoration that I received on 19 August is greater and will do me more honour" [26].

Henriette, who was obviously delighted with the good news, gave Antoinette the present that she had promised if her father succeeded: permission to organize a baptism ceremony for her doll [27], whose godmother would be Sainte-Claire Deville's daughter!

At that time, Janssen was at Calcutta, where he stayed for a month to get used to the cold before leaving for the Himalaya: "it will be really nippy up the mountain" [28] he had been told. While there, he took a bit more time to write, even sending his cousin Geneviève Élisa accounts of the region that he was visiting:

> India is now a curious country, especially from the point of view of races and religions. Studies that have been made are, unfortunately, very incomplete; that of the monuments is still to be carried out. The English are only concerned with making money, and the other peoples are indifferent. [29]

The Stay in the Himalaya

There were no echoes of the meeting of 26 October at the Académie des sciences until he was at Simla, at an altitude of 8,000 ft (2,400 m), and it was doubtless, thanks to Faye's suggestion that he felt no animosity toward the "English scholar," whom he did not know, and whom Henriette spoke about to him without even giving him the name. As such, he wrote, on 12 December, to the permanent secretary of the Académie des sciences, Jean-Baptiste Dumas:

> I cannot accept the far too flattering praise that M. Faye has heaped on the results of my efforts, but I fully associate myself with that illustrious astronomer in applauding the success of M. Norman Lockyer. Given his ignorance of the results that I had obtained in India, this physicist fully deserved his achievement in obtaining, in an independent fashion, confirmation of his very sound predictions. [30]

He was not as pleasant to Henriette, who complained about his return being deferred from month to month, and to whom he replied:

> You can trust me; I have started to acquire glory and position for you, you have all you need, you have an interesting child to raise, you can do good, etc., and still you complain! But that is to tempt God. Do you want everything without any sacrifice? I beg you, once for all, show more courage and await my return with cheerfulness, work and spirit. [31]

Others were taken to task, such as Le Verrier, which was not surprising, but so was Rayet:

> The observation by M. Rayet is extremely mediocre; he did not even see that it was hydrogen, and reported B for C. His cross-shaped slit with the solar crescent was a poor arrangement, etc. M. Leverrier *[sic]* got him a decoration because I had one, through jealousy – that man is pitiable. [32]

For Janssen, one good piece of news brought others: he was proposed for the Académie des sciences on 7 December, which was an honor, even if the chances of success were non-existent on that occasion and, on 5 January Henriette learned that the major astronomy prize of the Académie des sciences, the Lalande Prize, was going to be awarded to him [33]. What is more, the Académie des sciences would increase the sum from 600 F to 2,000 F![4] [34] Henriette, who, since 16 June 1868 [35], had power of attorney over all the marriage's assets, was thus able to pay her next month's rent (225 F per month) without having to borrow from her brother …

On his mountain, Janssen worked "under conditions of atmospheric dryness that were exceptionally favorable for his spectral and hygroscopic research," and rejoiced in them: "I have superb weather and am gaining a marvelous advantage over M. Lockyer, who must now be under a curtain of fog" [36]. However, he allowed himself one unexpected distraction:

> I am on what is known as the Mountain of the Monkeys. There's a small female monkey that amuses me with her tricks and who likes me. During the day she plays in a large oak that is opposite my door. She is a quadrumanous monkey with marvelously formed hands. And so nimble! And what mischievousness! This is my main distraction. [37]

In a letter of 8 January, he announces the discovery of "prominence material around the whole of the solar disk, where it makes a form of continuous ring of which the prominences are just the most salient portions" [38]. He arrived at this result by using colored glasses or else opaque or semi-transparent diaphragms, and by placing the slit of his spectroscope, not (as formerly) perpendicular to the prominences, but tangential to the disk of the Sun [39]. However, when he prudently sent confirmation of the new discovery by telegraph on 12 January: "Confirmation of the existence of a hydrogenous atmosphere around the Sun. Relationship between the presence of spots and prominences" [40], he did not suspect that he had been anticipated by Lockyer, and which he willingly acknowledged subsequently:

> M. Lockyer had noted the presence of this atmosphere back in November. I anticipated him by two months in discovering the method; he has anticipated me by one month in finding the Sun's hydrogen atmosphere, which he has called the chromosphere [41, 42].

Curiously, it was between Huggins and Lockyer that a quarrel arose [43] because Huggins claimed priority for the idea of the famous "method"! For Janssen and Lockyer, however, the events of 1868 marked the beginning of a deep friendship, undoubtedly encouraged by the unorthodox path that each had followed (Lockyer began his career as an amateur astronomer while he was employed by the War Office), and sealed forever by the fine medal that the Académie des sciences had struck a few years later, in 1872. This medal has the profiles of the two scientists on the obverse, while on the reverse, Apollo's chariot passes in front of the eclipsed Sun. The god is pointing to the prominences with his right index finger, and the date of the 18 August is also stated on the reverse (Fig. 4.3). This annoyed Lockyer slightly, but Janssen explained that it was Jean-Baptiste Dumas, the permanent secretary to the Académie des sciences who had the idea of the medal, and that he

[4] In fact, 2,500 F would be awarded to Janssen at the meeting of 7 June.

Fig. 4.3 *The Janssen–Lockyer Medal: the profiles of the two scientists and Apollo's chariot passing in front of the Sun (top and bottom).* © Author's collection

wanted to commemorate the fact that it was the eclipse that led to the discovery [44]. When Janssen sent the medal to Lockyer, it was with the following words: "English science has always shown great sympathy towards me of which I am proud; please receive in return this evidence of the great esteem that we have of you in France" [45]. Lockyer hurried to express the pleasure that he felt in the fact that "the work it commemorates has been accomplished with such a noble fellow as yourself" [46].

Janssen was keen to extend his stay in High Asia to study not just the spectra of the planets and stars, searching for water vapor (which he found), but also the higher reaches of our own atmosphere. But funds were lacking. He wrote to Duruy, who hastened to help him, whereas Faye and Dumas thought that his presence in Paris would be more useful. On 30 January, the minister wrote to Dumas: "The results of this expedition are assuming such importance that it seems to me to be urgent to authorize M. Janssen to continue. I am writing to him by telegraph to continue his research and that the necessary funds will be placed at his disposal" [47]. On 6 February, 5,000 F were released!

Henriette met the minister to receive this money, and in April Faye confided in her:

> Before this expedition, M. Janssen was known as a talented person, and highly intelligent, but now, he has made a name for himself, he is perfectly set. He has, he added laughing, got things in a bag. His observations of the eclipse succeeded perfectly, whereas the expedition to Malacca has returned sheepishly; the English themselves achieved very little; and when he made his discovery he far exceeded the aims of his expedition. [48]

Later, Janssen would never forget the effective support from the minister: "We, the French observers, must recognize that to M. Duruy we owe the fact that we have had favourable conditions to sustain the scientific struggle that arose between scholarly nations regarding the great astronomical phenomenon of 1868" [49].

The Principle of the Spectrohelioscope

Janssen did not return to France until the end of July, and in August he did not travel to the United States to experience the eclipse during which the American astronomers Young and Harkness observed, for the first time, in emission, an unknown green line in the corona (then christened "line 1474" because of its location on Kirchhoff's scale, and subsequently, in 1882, the "coronium line").[5] At the time he was in England, at Exeter, with Henriette and Antoinette. The annual meeting of the BAAS was being held there, and Janssen intended to describe his recent discoveries. These included a short paper that was slightly more detailed than the one presented to the Académie des sciences on 11 January [50], on a "method of observing monochromatic images of luminous objects", which, "applied to the Sun, can provide images of the whole of prominences" [51]. To do this, Janssen suggested placing a second slit at the exit of the spectroscope, on the monochromatic image of the entrance slit,

[5] We now know that it is a line emitted by an atom of iron, 13 times ionized (Fe XIV).

where the image of the Sun was formed. "If we now imagine that the spectroscope turning around an axis through the two slits, then the various portions of the luminous image *[of the Sun]* will successively produce their monochromatic line in the examining telescope, and if the relative movement is sufficiently rapid, the succession of all these lines will produce an overall impression that will be the image *[of the Sun]*, created by rays of a single refrangibility" [52]. In other words, if the motion is sufficiently fast to obtain persistence of vision of the impressions on the retina, the trick will work! This is nothing less than the description of the principle of the spectroheliograph, which was perfected at the beginning of the 1890s by Hale in the United States and Deslandres in Paris, and interest in which has never flagged. But it would have been obvious that Janssen's suggestion was directed at colleagues who would be expected to understand, and not to the general public.

At Exeter, Janssen became acquainted with Lockyer, and the two families subsequently met again in London.

Janssen and the Helium that He Did Not Discover

Janssen is frequently attributed with the discovery of helium during the course of this eclipse. It was not so, but what role did Janssen play in the affair? In all probability, it was simply that of a mere expert in "observational skill."

In fact, on 18 August, all the observers, including Janssen, mentioned a yellow line observed in the spectrum of the prominences, but no one was able to give the exact position of the line. It was Lockyer who really became concerned about it after 20 October, thinking about it calmly in London, and after having read the report from his colleagues, and from whom he had probably requested further information. He established positively that this line did not coincide with the D_1 and D_2 sodium doublet. The line, which he called D_3 to distinguish it, was "more refrangible," which means that it had a shorter wavelength than the other two. It seems obvious that Janssen was replying to a specific query that he had received, when he asked Charles Sainte-Claire Deville to bring the attention of the Académie des sciences the following passage from his letter of 19 December 1868 (which has never been published): "Several observers have claimed the bright D line as forming part of the spectrum of the prominences on 18 August. The bright yellow line did indeed lie very close to D, but the light was more refrangible than those of the D lines. My subsequent studies of the Sun have shown the accuracy of what I state here" [53]. It also appears perfectly clear that it was thanks to his observations made after the eclipse that he could be so sure of what he stated, but he never returned to the subject.

Lockyer, for his part, wanted to identify the line and, in collaboration with the chemist Edward Frankland, tried to observe it in the laboratory, thinking at first that it was a hydrogen line that was not emitted under the same conditions as the C and F lines. In view of the total failure of the experiments, Lockyer decided to christen the unknown solar element that emitted the line "helium."

It was only in 1895 that helium was extracted from a mineral, cleveite, by William Ramsay, and was thus identified as being a terrestrial element.

Lockyer did not mention Janssen in the article on the history of helium that he published in 1896 [54] in his journal *Nature*,[6] and Janssen certainly never claimed the discovery of this gas, in which he appeared so disinterested that one might wonder if he ever spoke its name. He did, and three times at least: the first under indisputable circumstances in 1896, because it was on the occasion of the award of the Académie des sciences Janssen Prize to Deslandres, who had indeed specifically identified in cleveite two other solar lines caused by helium [55]; the second citation occurred during the course of a public lecture on the constitution of the Sun and on the Mont-Blanc Observatory, which Janssen gave in Rouen in 1901 [56]; and finally, helium is definitely given due prominence in the account of the eclipse of 1905.

References

1. Letter from Jules to Henriette Janssen, 10 October 1868, BIF Ms 4133–106.
2. Janssen, Jules, "Conférence sur l'éclipse totale du 18 août 1868" *[lecture given in 1870]*, in *Lectures académiques discours*, Paris, Hachette, 1903, 299–329, p. 304.
3. *Ibid*., p. 135.
4. Aubin, David, "La métamorphose des éclipses de Soleil", *La Recherche*, No. 321, June 1999, 78–83.
5. Ministerial order, 09 March 1868, AN F[17] 2977.
6. Letter from Jules Janssen to Urbain Le Verrier, 12 March 1868, Marseille Observatory archives (AM).
7. Letter from Victor Duruy to Urbain Le Verrier, 19 March 1868, AM.
8. Cited in Yvert, Benoit (ed.) *Dictionnaire des ministres (1786–1989)*, Paris, Perrin, 1990, p. 252.
9. Georgelin, Yvon and Arzano, Simone, "L'éclipse du Soleil du 18 août 1868, Stephan and Rayet, hôtes du roi de Siam à Wha-Thonne", *l'Astronomie*, **113** (1999), 12–17.
10. Janssen, Jules, "Conférence sur l'éclipse totale du 18 août 1868" *[lecture given in 1870]*, in *Lectures académiques discours*, 299–329, p. 306.
11. Letter from Jules to Henriette Janssen, 04 August 1868, which arrived in Paris 02 September 1868, BIF Ms 4133–103.
12. Janssen, Jules, "Observations spectrales prises pendant l'éclipse du 18 août 1868, et méthode d'observation des protubérances en dehors des éclipses", *CRAS*, **68** (15 February 1869), 367–76.
13. Janssen, Jules, "Indication de quelques uns des résultats obtenus à Cocanada *[sic, instead of Guntoor]*, pendant l'éclipse du mois d'août dernier, et à la suite de cette éclipse", *CRAS*, **67** (26 October 1868), 838–9.
14. Janssen, Jules, "Sur l'étude spectrale des protubérances solaires", *CRAS*, **68** (11 January 1869), 93–5.
15. Janssen, Jules, "Indication de quelques uns des résultats obtenus à Cocanada *[sic, instead of Guntoor]*, pendant l'éclipse du mois d'août dernier, et à la suite de cette éclipse", *CRAS*, **67** (26 October 1868), 838–9.
16. Janssen, Jules, "Conférence sur l'éclipse totale du 18 août 1868", in *Lectures académiques discours*, Paris, Hachette, 1903, 299–329, p. 316–7.
17. Janssen, Jules, "Observations spectrales prises pendant l'éclipse du 18 août 1868, et méthode d'observation des protubérances en dehors des éclipses", *CRAS*, **68** (15 February 1869), 367–76.
18. Telegram from Jules to Henriette Janssen, 19 June 1868, BIF Ms 4133–104.
19. Letter from Warren De La Rue to Henriette Janssen, 26 August 1868, BIF Ms 4136–337.
20. Letter from Jules to Henriette Janssen, sent 19 September 1868, arriving in Paris 24 October, BIF Ms 4133–105.

[6] He created the journal in 1869 and supervised it until his death.

21. Letter from Jules to Henriette Janssen, sent 10 October 1868, arriving in Paris 09 November, BIF Ms 4133–106.
22. De La Rue, Warren, "Sur une méthode employée par M. Lockyer pour observer en temps ordinaire le spectre des protubérances signalées dans les éclipses du Soleil", *CRAS*, **67** (26 October 1868), 836–8.
23. Lockyer, Norman, "Spectroscopic observations of the Sun", *Proceedings of the Royal Society*, **15** (15 November 1866), 256–8.
24. Faye, Hervé, *CRAS*, **67** (26 October 1868), 840–1.
25. Letter from Henriette to Jules Janssen, 26 October 1868, BIF Ms 4134–69.
26. Letter from Jules to Henriette Janssen, 10 October 1868, BIF Ms 4133–106.
27. Letter from Henriette to Jules Janssen, 26 October 1868, BIF Ms 4134–69.
28. Letter from Jules to Henriette Janssen, 22 October 1868, BIF Ms 4133–107.
29. Letter from Jules Janssen to Geneviève Élisa Janssen-Le Roux, 25 October 1868, archives of Nicole Worms de Romilly and Françoise Andlauer.
30. Janssen, Jules, "Sur l'étude spectrale des protubérances solaires", *CRAS*, **68** (11 January 1869), 93–5.
31. Letter from Jules to Henriette Janssen, 02 December 1868, BIF Ms 4133–111.
32. *Ibid.*
33. Letter from Henriette to Jules Janssen, 05 January 1869, BIF Ms 4134–85.
34. Letter from Henriette to Jules Janssen, 05 January 1869, BIF Ms 4134–86.
35. Power of attorney dated 16 June 1868, AN, MC/ET/CVII/955.
36. Letter from Jules to Henriette Janssen, 16 January 1869, BIF Ms 4133–116.
37. *Ibid.*
38. Janssen, Jules, "Résumé des notions acquises sur la constitution du Soleil", *CRAS*, **68** (08 February 1869), 312–4.
39. *Ibid.*
40. Faye, Hervé, "Note sur un télégramme et sur une lettre de M. Janssen", *CRAS*, **68** (18 January 1869), 112–4.
41. Lockyer, Norman, "Spectroscopic observations of the Sun", *Phil. Trans.*, **159** (1869), 425–44, p.430.
42. Janssen, Jules, "Conférence sur l'éclipse totale du 18 août 1868", in *Lectures académiques discours*, Paris, Hachette, 1903, 299–329, p. 320.
43. Meadows, Arthur J., *Science and Controversy. A biography of Sir Norman Lockyer*, Macmillan, 1972, p. 55.
44. Letter from Jules Janssen to Norman Lockyer, 19 April 1872, Exeter University Library (EUL).
45. Letter from Jules Janssen to Norman Lockyer, 28 April 1872, EUL.
46. Letter from Norman Lockyer to Jules Janssen, 13 May 1872, BIF Ms 4136–383.
47. Letter from Victor Duruy to Jean-Baptiste Dumas, 30 January 1869, AN F^{17} 2877.
48. Letter from Henriette to Jules Janssen, 04 April 1869, BIF Ms 4134–100.
49. Janssen, Jules, "Conférence sur l'éclipse totale du 18 août 1868", in *Lectures académiques discours*, Paris, Hachette, 1903, 299–329, p. 299.
50. Janssen, Jules, "Sur l'étude spectrale des protubérances solaires", *CRAS*, **68** (11 January 1869), 93–5.
51. Janssen, Jules "Méthode pour obtenir les images monochromatiques des corps lumineux", in *Report of the 39th Meeting of the BAAS, Exeter, August 1869*, 67–8.
52. *Ibid.*
53. Janssen, Jules, "De l'existence d'une atmosphère hydrogénée autour du Soleil", *CRAS*, **68** (18 January 1869), 112
54. Lockyer, J. Norman, "The Story of Helium", *Nature*, **53** (1896), 319–22 and 342–6.
55. Janssen, Jules, "Rapport sur le prix Janssen décerné en 1896 par l'Académie des sciences à M. Deslandres", *CRAS*, **123** (21 December 1896), 1120–1.
56. Janssen, Jules, "La constitution du Soleil et l'observatoire du mont Blanc", in *Œuvres scientifiques*, vol. II, 5580–66.

Chapter 5
The Eclipse of 1870, Balloons, and Patriotic Missions

> *"Le XXe siècle [...] verra réalisées les grandes applications de la navigation aérienne, et l'atmosphère terrestre sillonnée par des appareils qui en prendront définitivement possession."*
>
> *[The 20th century [...] will see the realization of major applications of aerial navigation, and the terrestrial atmosphere will be furrowed by devices that will take complete possession of it.]*
>
> Jules Janssen [1]

In 1868, the spectral analysis of prominences had monopolized his efforts and, as Janssen stated in 1870, "the magnificent halo of light that surrounds the eclipsed Sun, and which is generally termed the aureole [...] was, of necessity, neglected and reserved for later study" [2].

Departure from Besieged Paris on Board the Balloon *Le Volta*

The first favorable occasion that would enable him to devote himself to this issue was the total eclipse of 22 December 1870, visible in the Mediterranean basin. Janssen decided to travel to Oran to observe it. But how could he get through the Prussian lines that were then encircling Paris? His British colleagues made every effort and obtained a special safe-conduct from Bismarck, but Janssen was too much of a patriot to make use of it, especially as he was responsible for transmitting a verbal message to Gambetta, the Minister of the Interior and acting Minister of War, who had rejoined the "délégation de Tours"[1] by balloon the preceding 7 October. The only solution was the same: an aeronautical trip. At the request of the Académie

[1] Members of the acting government, who had withdrawn to Tours (235 km south-west of Paris).

F. Launay, *The Astronomer Jules Janssen: A Globetrotter of Celestial Physics*, Astrophysics and Space Science Library 380, DOI 10.1007/978-1-4614-0697-6_5,

Fig. 5.1 *Janssen taking off from besieged Paris on board* the Volta, *2 December 1870. Tablet at the base of the statue.* © Paris Observatory Library

des sciences, the Minister for Public Education, Worship and Fine Arts, Jules Simon, entrusted the mission to Janssen, had 5,000 F allocated to him, taken from the residue of the Montyon Fund (the founder, among other schemes, of the "prix de vertu" [*virtue*] award given by the Institut de France) and, as a bonus, offered him the balloon! Janssen had no experience in the matter, but that didn't put him off:

> I had never made a free ascent, and for some time Paris had no longer had any experienced aeronaut to send to the provinces, but I did not feel I should let myself be stopped by this difficulty, and, being convinced that theoretical knowledge, carefully acquired, and experience in travel would suffice to give me the confidence and the necessary inspiration to control my aerostat properly, I undertook its supervision. I believe that the results have shown that I was right. [3]

Janssen took off, then, from the Godard workshop at the Gare d'Orléans (now the Gare d'Austerlitz) on 2 December 1870 at 6 o'clock in the morning (Fig. 5.1). He was on board the *Volta*, with a single companion, a sailor named Chapelain.

The permanent secretary of the Académie des sciences, Jean-Baptiste Dumas witnessed the departure, together with his son-in-law Hervé Mangon, director of the Conservatoire des arts et métiers, Charles Sainte-Claire Deville, and other notables.

> The signal is given, and the balloon rises slowly [...]. However, the dawn was beginning to break and was already coloring the area to the east with a whitish tint that was growing rapidly. But this glimmer existed only for us: Paris was still in darkness and was visible only through punctuated lines of fiery light that picked out its major arteries.

> The contrasting impressions that the lower regions presented relative to ours were striking: below our feet, below a heavy, dark atmosphere, were the night-time trappings of a great city, whose reddish and volcanic fires kindled the idea of a lower world with its appetites, passions, violence, and misery. And what a coincidence! Was Paris not at that very moment struggling against the fierce grip of its enemies. But if, breaking away from these ideas, we returned our view to our pure, translucent region, already flooded with the light of dawn, what a contrast and what an enormous relief! One felt lightened, and filled with a feeling of indefinable purity which gently led thoughts to an extra-terrestrial order of ideas. But it was necessary to avoid these impressions and think of controlling the balloon. [4]

The flight was carried out without any difficulty at all, and the aerostat, which had risen to about 2,000 m, descended at the end of the morning at a small village in the Saint-Nazaire district, without injuries or breakages, and despite the high winds that were then raging. The travelers had covered 400 km in 5¼ h. "That was an average speed of 76 km, or nearly 20 leagues, per hour" [5].

> Meanwhile, farmers arrived from all directions, and in an instant we were in the middle of a crowd that pressed in on and suffocated us. These nice people had never seen a balloon. They overwhelmed us with questions: [...] Monsieur, you have come from Paris, is it suffering a lot; have they got food supplies for long? [6]

Janssen and his companion had to answer the patriotic anxiety of the local inhabitants, with no interruption other than a lunch that was greatly appreciated by the Parisians, because it included eggs, butter, and chicken, before they were escorted to the station, where the balloon and its precious cargo were awaiting them.

The Mission to the Government

A special train then took Janssen to Nantes from whence he reached Tours. It was 11 in the evening when he arrived and was able to deliver his message to Gambetta. The existence of this diplomatic mission was obviously not revealed immediately, and this is why, much later, Janssen's friends on the other side of the Channel, and Lockyer in particular, never knew if the Prussian safe-conduct had reached him. Indeed, it was only when he came to write Janssen's obituary for the Royal Society (the British equivalent of the Académie des sciences) that Sir Norman wrote to Madame Janssen to ask her for details, and received the following reply from her (Fig. 5.2):

> Monsieur Janssen knew of the steps taken by your government to obtain free passage for him from M. de Bismarck but, morally, he could not profit from it; having been entrusted by General Trochu and Jules Simon with a verbal assignment to contact Gambetta at Tours, where the Government of National Defense was located, he hastened to leave without awaiting the reply from M. de Bismarck. But he was, believe me, exceptionally grateful to the British Government, whose esteem and generosity he always appreciated, as well as to you, who took such a large part in the matter. As soon as he arrived at Tours, he went to see Gambetta before traveling to Algeria, and for a number of years it was not possible for the particular circumstances of his departure to be made known. [7]

Fig. 5.2 *Jules Simon (left) and Léon Gambetta (right)*. Album Félix Potin. © Author's collection

The day after the flight, before leaving for Marseille via Bordeaux, and then for Oran, Janssen met Adolphe Thiers. "This visit," he said "was the beginning of the extremely pleasant and precious relationship that I had with this great statesman" [8].

The Scientific Mission and the Invention of the Aeronautical Compass

Inflated by lighting gas, like all the balloons used during the siege of Paris, *Le Volta*, which had a capacity of 2,000 m^3, had a lifting power of about 1,400 kg. Taking account of the weight of the balloon and of the basket (520 kg), that of the ballast (570 kg) and of the travelers (150 kg), the weight of any instruments could not exceed 160 kg, "a difficulty that appeared insurmountable in the eyes of some members of official circles" [9], as Janssen explained. To him, as we know, nothing was insurmountable. All that was necessary was to get himself appropriately organized. So Janssen carried just the essential elements of his instruments, "intending to complete them in a major city." He placed one instrument per case, and each case was padded externally such as to be able to withstand a fall of some 10 m onto the ground without damage. And this is how he transported, apart from tools, a 37-cm aperture reflector, a 16-cm reflector, a 108-mm refractor, and a collection of spectroscopes, polarimeters, and other barometers.

At Oran, everything was ready to study the solar halo, but the overcast sky at the time of the eclipse unfortunately prevented any observations whatsoever, neither for Janssen nor for the members of the British expedition who had chosen the same site:

Huggins, Tyndall, and Admiral Ommaney. Lockyer had no better luck in Sicily, where he had the additional misfortune of suffering shipwreck [10]! The American, Young, made a particularly significant observation, however, because he observed a reversal of the spectrum at the base of the chromosphere [11], which confirmed Kirchhoff's theory.

Nevertheless, Janssen had not wasted his time as far as scientific matters were concerned because during the flight he conceived a new instrument: the "aeronautical compass," designed to give the direction and speed of aerostats. "Struck by the accidents and even tragedies that have occurred as a result of the ignorance of the track followed by the aerostat, I attempted to fill this lacuna" [12] Janssen wrote simply to Dumas on 19 February 1871, from Bordeaux.

The Stay at Bordeaux

After his return from Algeria, Janssen did not, in fact, return to Paris. From Marseille, he traveled to Bordeaux, where he found the provisional government, which had left Tours. Once there, his stay became prolonged.

Once again, Henriette was separated from her husband and, at that period, the circumstances were particularly dramatic. At the beginning of the Franco-Prussian War, in August 1870, the Janssens had entrusted their daughter Antoinette to Henriette's sister, Caroline Blémont, who had withdrawn with her daughters to Le Havre and her brother's house. Henriette, who was then 42 years old, was pregnant again and, as with her first pregnancy, had a stillborn child the following 12 September [13]. After her husband's departure, she thus remained in Paris with the sole family member being her 81-year-old mother-in-law, and any news took a long time to arrive – if it arrived at all. In fact, the "pigeon telegraph" that Janssen used was far from functioning perfectly. It was perhaps worse when news arrived distorted: although on the 16th it was announced that the *Volta* had landed at Saint-Nazaire without accident, the paper *Le Temps* published a completely inaccurate account on the 29th of a meeting at the Académie des sciences where it had been stated that all Janssen's instruments had been broken in landing. Henriette was shaken by the thought of the shock that Janssen must have sustained. In addition, in Paris, where a smallpox epidemic was spreading, the cold intensified (the ground was frozen to a depth of 50 cm after 21 December), and the bombardments started on 5 January 1871. It was the very next day, about 4 or 6 o'clock in the morning, that "in her sleep and without giving the slightest sign" [14] Janssen's mother expired peacefully. Luckily, Henriette still had her maid to help her, and the cousin Eugène Janssen was able to deal with the burial at the Montmartre cemetery. Henriette wrote letter after letter to her husband. The missives left by balloon, but more than half of them never arrived. Henriette, for her part, received no news from her husband until 7 February, when he was in Bordeaux. Bound up in all sorts of tasks, Jules obviously did not understand the situation in the capital, nor the difficulty of communications because he dared to complain about not receiving enough letters, and too many letters that consisted of just a single page! He demanded details on all sorts of matters while Henriette continued to repeat at length the sad news, both personal and about the Siege, while his letters to her remained laconic. He even refused to give his address. Although officially occupied with balloons, there was no doubt that he was also playing a significant role for the government, as during the months before his departure. When his statue was dedicated, Prince Roland Bonaparte did indeed declare: "During the terrible year, he devoted himself to creating and supervising military observation posts" [15]. On 1 February, Janssen confided in Henriette:

> I am currently at Bordeaux where I am of use: I shall stay here until the end of events. You know that they are going to convene a National Assembly here. I have struck up a friendship with M. Thiers. I visit him in the evenings, as a family friend: we talk about science, history, and art. I revel in the marvelous organization *[of his mind]*, which tackles everything with such a rapid and sure comprehension. He knew my work, and welcomed me with particular distinction. [16]

Things became a little clearer somewhat later: "It may become important for me to be here where the government is" [17] he wrote on 20 February.

Henriette was prepared to accept the situation, provided that her husband should make himself clear, and that he should allow her to join him at Bordeaux, or to return to her family and Antoinette at Le Havre; but it seemed that he wanted neither one nor the other. Henriette bravely concerned herself with the little female monkey that had been brought back from India, who behaved herself perfectly, but she herself became more and more unwell. Given the circumstances, it was enough to depress anyone. On 1 March, which she called "a day of mourning" because it was the day on which the capitulation was ratified by the Assemblée Nationale, she could not stand it any longer:

> If you go to England or if your duty retains you at Bordeaux, I shall come to join you if you call; if not I shall leave for Le Havre. Tell me clearly which plan I should adopt, but don't lose sight of the fact that I cannot remain alone any longer. Take account of the difference between our two natures; I am a woman, and perhaps, possibly, even a weak woman in every sense of the word, but the day that I cease to live from the heart, everything within me will be shattered. I want to be loved, and everything I do is for that, and I want to live with those who love me. I see human existence now as being too short to think and act differently. – As for our unhappy country, let M. Thiers return from Bordeaux as quickly as possible, because his presence ought to deliver us from our base and harsh enemies. Paris is in mourning just as long as their feet trample our soil, which is soiled by their presence. [18]

It was only on the 2 March that Janssen finally said that he was living at the state primary school. His departure had been decided: "Peace having been made (voted for yesterday), I have no more need to be here, and I shall not leave you alone, my dear plaintiff, any longer than I need to complete my duties" [19]. After a brief encounter between the couple during March, on the 25th, Henriette was at Le Havre, where she recovered, and Jules, who was "extremely vigorous," whereas he had been unwell at Bordeaux, had left for London on a new mission, which was to last more than a month, and a few additional days in July! He was officially charged by the minister to study the astronomical institutions in England [20], and he certainly intended, once there, to develop greatly his relationships with English colleagues, and to become acquainted with the Royal Society.

While Paris lived through difficult times under the Commune, Janssen was indeed well received in London. He was impressed by a meeting of the Royal Society which had "a lot of style. The President in a monumental chair surmounted by the arms of England, a great silver mace on a long velvet cushion, etc., and the portrait of Newton above the President" [21]. He not only met many "scholars," who generally extended an invitation to him: Warren De La Rue, Tyndall, Wheatstone, the chemist Williamson ("Yesterday I dined at Mr Williamson's. His daughter is very pretty, brunette, and about the same height at Antoinette. – Madame is tall and good-looking. – Great luxury, as always" [22]), Lockyer, General Sabine (President of the Royal Society), Admiral Ommaney, but also the widow of Admiral Manners, who was none other than the daughter of the Marquis of Noé. She gave him a letter of introduction to her brother, Count Amédée de Noé, who was the celebrated humorist Cham, whom Janssen wanted to know "because of his delightful sense of criticism" [23]. He also went to Tulse Hill, the private observatory of Huggins, who was the same age as him, but who had yet to marry: "Two days ago I visited M. Huggins 4 miles from London. He is a bachelor, and lives in a delightful house full of objets d'art. I saw the refractor that was built at the Royal Society's expense; it is magnificent" [24].

Janssen, who represented France, took his role very seriously:

> Because my trip is truly a scientific mission. I am taking notes on the establishments, the instruments, the work, etc., and if, in the work of reorganizing France, anyone should consider consulting me, I would be able to give good advice. Yesterday I was at the astronomical society's dinner,[2] with MM. De La Rue and Lockyer. There were speeches. I stressed the necessity for England to see the Government intervene in the matter of science and to support scientists. I was extremely emphatic about this, and the wish was expressed that I would give my advice to the Commission that the Government has established to carry out an enquiry on this subject. After the dinner we went to the Society's main building, where I saw the principal London astronomers. The President informed the meeting of my presence, and everyone tapped their umbrellas on the floor as a sort of greeting to the Queen. I was invited to speak and I spoke about the next eclipse and the arrangements that should be made. [25]

Janssen was, of course, on a scientific mission, but that did not prevent him from attending a concert, ensuring, however, that he had the benefit of a half-price seat, an onerous condition that was forced on men of science:

> As I could not hear that beautiful music with you, my dear companion and friend, I am writing to you with my head still full of the beautiful motifs. I have just heard *Les Huguenots* with Patti. It only cost me 2 schillings *[sic]* and six pence, because I went to the hall, which is a 5-shilling one, and I cleverly profited from the English custom of only paying half-price once the curtain has risen. – See to what men of science are reduced – who are said to bring honor to one's country – in the year of grace, or rather disgrace, 1871! – We are the ones who should establish the Commune and erect barricades against a blind and ungrateful Society. But let's forget all that. Our reward is found within ourselves [26].

As far as the performance was concerned, he added:

> I return to Patti, whom I had never heard, but whom we shall hear – you understand me? – although this singer is not to my taste (far from it). I find that she has a great voice, especially in the upper register where she produces notes of great brilliance and power, but she

[2] Of the RAS Club, a select group of Fellows, not a formal dinner of the whole Society.

does not have a very great range, from what I could tell from hearing her just once. Above all, however, I would condemn her for lacking in feeling – what one might call soul – which should always dominate and guide the performance, and without which no one is a great artist. – Dupré *[sic, instead of Duprez]* was a great artist, because science, art, and feeling never left him. – Patti triumphantly projects an exemplary singing exercise, where she performs material of great vocal difficulty, but it is more of a physical triumph, which leaves me rather cold. – But the surroundings are so fine, the singers around her so good, and the orchestra is so well directed in support of her and to increase the esteem in which she is held, that, overall, such an evening was the greatest pleasure!

Janssen Awarded a Doctorate at Edinburgh

Janssen crossed the Channel again in July, when he attended the BAAS meeting in Edinburgh, where he was to be made Doctor of Laws on 1 August.

That's the ceremony over. I put on the gown this morning. We were placed around the Lord Justice General on a dais – in front there was a fine audience of students, ladies, etc. An account of my work was read out in public – Santorin, atmospheric lines, the Faulhorn, Lake Geneva, the experiments at La Villette, and the prominence method in India. And then the Siege of Paris and my departure in the balloon. – All this with great praise, all fiercely applauded – especially when it came to the balloon, when the applause lasted several minutes. After this eulogy, the Chancellor placed the cap on my head and then I was invested with a sort of stole in blue silk, and I sat down. So I am now honorary doctor of the University of Edinburgh, of which the great Brewster was Principal at the end of his life. A number of the most eminent British scholars were honored at the same time as me, notably Huggins and also Joule, one of the proponents of the mechanical theory of heat. [27]

That same evening, more good news arrived from the Royal Astronomical Society:

M. Pritchard has informed me that the Astronomical Society has nominated me as a Foreign Associate at a first meeting. There has to be a second, and then I will be sent the certificate. That will take time, but it will be a considerable achievement, because there are just one or two French astronomers who have been accorded this honor. [28]

Two days later, Janssen gave an account of his balloon voyage. Among the audience was Dom Pedro II, the Emperor of Brazil, a great enthusiast of science, whose acquaintance Janssen made. Janssen was more than satisfied by his trip:

I am about the only scientist representing France at the meeting. I have been received and congratulated in an extremely friendly manner. [...] Overall, I am very pleased; I have arrived at the end I hoped to achieve: representing my country and getting myself accepted by foreigners. [29]

Janssen and Aeronautics

The escape from Paris was the only balloon flight that Janssen made, but for him, this epic, which so impressed the masses, represented the beginning of a sustained and visionary interest in aeronautics.

In 1873, when he was the first chairman of the infant Société française de navigation aérienne, he represented the overall approval that aerostatics naturally acquired, which had been initiated "chez nous" *[i.e., in France]*, and which obviously developed during the Siege. But he noted that the aerostat and the researchers who wanted to use it had before them "an immense field of study, and a whole science to create. [...] The field is extremely rich, it is completely virgin territory, and the first to launch themselves into it, if they are educated, persistent, and courageous, will make major discoveries" [30].

Janssen suggested dividing the work of the society into three types of study: "First Aeronautics as applied to military purposes; Second Aeronautics applied to meteorology; and Third Aeronautics as such, and techniques, which might be linked to all the problems arising from the study of the flight of birds, the direction of flight, etc."

Very soon, although he did not "fly" himself, he commissioned aeronauts to carry out research for him. Thus on 22 March 1874, he asked Joseph Crocé-Spinelli, the inventor of the water velocipede and who suggested the variable-pitch propeller, to make spectroscopic observations of water vapor: "It is a question of knowing whether, in the upper regions above the observer's head, there still exists a significant quantity of water vapor" [31]. Janssen recommended observing the strongest lines in the spectrum. The balloon ascended to 7,500 m, and "the lines weakened as one ascended, and above 7,000 m [...] M. Crocé-Spinelli found that they disappeared completely from the spectrum, although the light was very strong, and the neighboring lines, notably F, were very clearly perceptible" [32]. Also on board *L'Étoile polaire* was Théodore Sivel, a former ship's navigator and the inventor of a [*nautical*] conical anchor and an improved guide-rope – a rope that is trailed on the ground when a balloon lands, to decrease its speed and assist the final descent – fitted with a device to increase drag. Regrettably, the two aeronauts died on 15 April 1875, on board the balloon *Le Zénith*, when only the third aeronaut, the chemist who founded the journal *La Nature*, Gaston Tissandier, was still alive when the landing took place some 250 km from Paris, as the crow flies. The maximum altitude of 8,601 m that the balloon reached, and which was fatal for the two men, was confirmed when closed and sealed boxes, containing "barometric witness tubes, devised by M. Janssen, and already used by Sivel and Crocé-Spinelli during their ascent to 7,500 m" [33] (in 1874), were subsequently opened in the laboratory. The tubes in question, 50-cm long, were originally filled with mercury. The clever part was that at their base there was a capillary section, curved upward and open at the tip, from which the mercury could escape as the pressure dropped during the ascent. On the return, all that was required was to create a (partial) vacuum in the laboratory[3] at the bottom, until the mercury, which had risen again when the basket returned to the ground, was just level with the end of the capillary. The tube was then in exactly the same state as it was when the maximum altitude was attained, which could thus be determined precisely.

Later, Janssen was in an ideal position at Meudon to follow the evolution of aerostatics, which had been carried out there ever since 1793 when, at the suggestion of Gaspard Monge, the Comité de salut public created in the "Maison Nationale de Meudon" (the château neuf), a national establishment where the first military

[3] In this case it was the laboratory of Marcelin Berthelot.

aerostat had been created [34]. In 1875, a "Committee for communications by aerial means" was established, with Colonel Laussedat as President and a graduate of the École Polytechnique, Captain Charles Renard of the Engineers as Secretary. The activity at Meudon picked up again because around the Chalais pond there were workshops available, which Renard requested should be allocated to the group. This happened in 1877, thanks to the support of Gambetta: the Central Establishment for Military Aerostation was born. There, Charles Renard, his brother Paul (also a graduate of the École Polytechnique), and the engineer Arthur Krebs, carried out their work with dedication and enthusiasm. In 1885, the large dirigible *La France* was completed after several modifications. On board, Charles Renard and Krebs made seven flights around a closed circuit, a milestone in the history of aerostatics, since the very first flight on 9 August 1884. Aware of the significance of the event, Janssen "arranged, at the Meudon Observatory *[which he directed]*, for the necessary arrangements to be made to obtain photographs. On 23 September 1885, as the balloon was returning to Chalais, a good image was obtained, which was used to make a photogravure plate," which he exhibited to the Académie des sciences [35].

In his speech at the opening of the International Congress of Aeronautics and Pigeon-fancying, which was held in Paris in July 1889, and of which he was chairman, Janssen, who had mentally outstripped closed-circuit flights, declared:

> So I do not hesitate to say that the 20th century, which we have almost reached, and whose dawn we are now welcoming, will see the realization of major applications of aerial navigation, and the Earth's atmosphere will be furrowed by devices that will take complete possession of it, either to study it daily and systematically, or to establish communication and links between nations, which will make light of continents, seas and oceans. [36]

In 1890, Janssen paid tribute to Gay-Lussac, "the first to really examine scientifically the upper regions of out atmosphere" [37], and the following year he advocated attempting to carry out observations from a balloon of the forthcoming transits of Mercury in front of the Sun [38].

At the final session of the meeting of scientific societies in 1892 at the Sorbonne, Janssen gave a major lecture in which he traced the history of aerostatics, and expressed the wish that the government and the whole of France should take an interest in aeronautical matters [39].

A new and very interesting scientific application of aeronautics came to light in 1898, at the time of the November Leonid meteor shower [40]. Janssen obtained the finances for a balloon flight to take place, with a young Russian astronomer, Alexis Hansky, then working at the Meudon Observatory, on board. He observed only 14 meteors, but Janssen fully intended to repeat the experiment the following year. The meteor shower had a maximum every 33 years, and this was in fact expected in 1899, so Janssen obtained the use of several places in two balloons. In the first, *L'Aéro-Club*, belonging to the club of the same name, were the Russian, Gabriel Tikhov, a student at Meudon Observatory, and Lespieau, a lecturer in chemistry at Chaptal College. On board the second, *Le Centaure*, owned by Maurice Mallet, a manufacturer and aeronaut, were Dorothea Klumpke,[4] an astronomer working at

[4] Later the wife of the famous astrophotographer Isaac Roberts.

Les étoiles profitent du vélocipède pour devenir toutes des étoiles filantes.

Fig. 5.3 *Shooting stars as seen by Cham.* Cours d'astronomie. © Author's collection

the Meudon Observatory, where she was in charge of calculations relating to the Carte du Ciel (whose director, Maurice Lœwy was happy to put her at Janssen's disposal), as well as the scientific journalist (yet also an aeronaut), Wilfrid de Fonvielle, who was responsible for noting down the observations. During the course of the two flights, which were perfectly carried out, the first team observed about 200 meteors between first and fourth magnitudes, whereas the second detected 19, which was very disappointing. A very detailed report was produced. In addition, the aeronauts dropped from the basket stamped postcards bearing the address of the observatory and the time that they were released. Anyone finding them was requested to return them after having added the location where they were found. "A number of these cards were returned to us" Janssen stated. "The information that they contained enabled us to reconstruct the path followed by the balloons and to determine the location of the observations and the corresponding times" [41]. The experiment was repeated in 1900, but because of atmospheric conditions, just two of the three flights envisaged were carried out, and merely four meteors were observed in total (Fig. 5.3)!

It was during the same period, in July 1898 and then in July and September 1899, that Count Aymar de La Baume Pluvinel, a wealthy amateur astronomer, who had been very seriously interested in astronomy since 1882, and who would be elected to the Académie des sciences in 1932, released several balloons carrying spectrographs to photograph the spectrum of the Sun [42] (Fig. 5.4). M. de La Baume had

Fig. 5.4 *Janssen at the (Botanical) Garden of Acclimatization on 18 July 1898 after the release of a balloon fitted with a spectrograph to study the Sun's spectrum. In front of him is Count Aymar de la Baume Pluvinel.* © Paris Observatory Library

put himself at Janssen's disposal from 1887 for the observation of several eclipses, and certainly Janssen would have followed operations very closely.

In September 1890, Janssen presided for the second time over the International Congress, which took place at Meudon. He recalled the history of balloons, mentioned their increasing importance for armies, and described the work that had been carried out at Meudon since 1886 on making balloons dirigible. He asserted loudly and strongly that "the nation that, in this respect [*aeronautics*] decides to make major advances, will gain a power and advantages, the results of which are impossible to predict today." The benefits were undeniable: increased knowledge of meteorology or better usage of sources of energy; however, he hoped that "these conquests, which assume an all-powerful industry and inspirational science, will signal the state of a civilization that is sufficiently advanced that it will recognize that the interests and the happiness of humanity are for justice, rights and peace" [43].

Fig. 5.5 *Janssen (sitting to the* left *of the Minister) at the official unveiling of Bartholdi's monument to the Siege aeronauts.* Le Petit Journal, *11 February 1906.* © Author's collection

In January 1906, Bartholdi's monument to the Siege aeronauts was unveiled at Neuilly. Janssen was one of the survivors among the 168 people who had crossed the German lines by air, between 23 September 1870 and 29 January 1871 and, of course, he gave a speech (Fig. 5.5).

References

1. Janssen, Jules, "Discours prononcé à la séance d'inauguration du congrès international aéronautique et colombophile le 31 juillet 1889", in *Annuaire du Bureau des longitudes (BdL) pour 1890*, 734–45.
2. Janssen, Jules, "Voyage aéronautique du Volta", in *Lectures académiques discours*, 3–20, p. 3.
3. *Ibid.*, p. 4.
4. *Ibid.*, p. 7–8.
5. Janssen, Jules, "Lettre à M. le secrétaire perpétuel, sur les résultats du voyage entrepris pour observer en Algérie l'éclipse du Soleil du 22 décembre 1870", *CRAS*, **72** (27 February 1871), 218–22.
6. Janssen, Jules, "Voyage aéronautique du Volta", in *Lectures académiques discours*, 3–20, p. 18.
7. Lockyer, Norman, "Obituary Notices of Fellows deceased, P.J.C. Janssen, 1824–1907, *Proceedings of the Royal Society of London*, **81** (1908–1909), lxxvii–lxxxi, p. lxxix.
8. Janssen, Jules, "Voyage aéronautique du Volta", p. 20.
9. Janssen, Jules, "Voyage aéronautique du Volta", p. 6.
10. Clerke, Agnes M., *A popular history of astronomy during the nineteenth century*, Edinburgh, Adam and Charles Black, 1885, p. 219.

11. Young, Charles, “Report in spectroscopic observations”, *Memoirs of the Royal Astronomical Society (MemRAS)*, 41, (1879), 432–7, p. 434.
12. Letter from Jules Janssen to Jean-Baptiste Dumas, 19 February 1871, Académie des sciences (AAS) archives, J.-B. Dumas archives, Janssen file.
13. Death certificate, Paris archives.
14. Letter from Henriette to Jules Janssen, 16 January 1871, BIF Ms 4134–110.
15. Bonaparte, Roland, in *Inauguration de la statue de Jules Janssen*, 25–9, p.26.
16. Letter from Jules to Henriette Janssen, 1 February 1871, BIF Ms 4133–134.
17. Letter from Jules to Henriette Janssen, 20 February 1871, BIF Ms 4133–136.
18. Letter from Henriette to Jules Janssen, 01 March 1871, BIF Ms 4134–119.
19. Letter from Jules to Henriette Janssen, 02 March 1871, BIF Ms 4133–138.
20. Letter from Jules Janssen to the minister, 22 July 1871, AN F^{17} 2977.
21. Letter from Jules to Henriette Janssen, 02 April 1871, BIF Ms 4133–141.
22. Letter from Jules to Henriette Janssen, 05 April 1871, BIF Ms 4133–142.
23. Letter from Jules to Henriette Janssen, April 1871, BIF Ms 4133–144.
24. Letter from Jules to Henriette Janssen, 05 April 1871, BIF Ms 4133–142.
25. Letter from Jules to Henriette Janssen, 15 April 1871, BIF Ms 4133–145.
26. Letter from Jules to Henriette Janssen, 15 April 1871, BIF Ms 4133–146.
27. Letter from Jules to Henriette Janssen, 01 August 1871, BIF Ms 4133–151.
28. Letter from Jules to Henriette Janssen, 02 August 1871, BIF Ms 4133–152.
29. Letter from Jules to Henriette Janssen, 03 August 1871, BIF Ms 4133–153.
30. Janssen, Jules, “Allocution prononcée à l’assemblée générale de la Société française de navigation aérienne le 26 novembre 1873”, *L’Aéronaute*, **7** (1874), 46–50.
31. Janssen, Jules, “Sur les expériences à effectuer pendant une ascension à grande hauteur”, *L’Aéronaute*, **7** (1874), 216.
32. Janssen, Jules, “Remarques sur le spectre de la vapeur d’eau, à l’occasion du voyage aérostatique de MM. Crocé-Spinelli et Sivel”, *CRAS*, **78** (13 April 1874), 995–8.
33. Tissandier, Gaston, “Mort de Crocé-Spinelli et Sivel”, *La Nature*, No. 100, 01 May 1875, 337–44.
34. Villadier, Francis and Marie-José, *L’aérostation à Meudon*, Ville de Meudon, 1983.
35. Janssen, Jules, “Présentation d’une héliogravure représentant les expériences aérostatiques de Chalais-Meudon”, *CRAS*, **102** (25 January 1886), 190.
36. Janssen, Jules, “Discours prononcé à la séance d’inauguration du congrès international aéronautique et colombophile le 31 juillet 1889”, in *Annuaire du BdL pour 1890*, 734–45.
37. Janssen, Jules, Letter as preface to *Gay-Lussac, aéronaute*, by Charles Sibillot, Orléans, 1890.
38. Janssen, Jules, “Sur le passage de Mercure”, *CRAS*, **112** (19 March 1891), 1098–9.
39. Janssen, Jules, “L’aéronautique”, in *Annuaire du BdL pour 1893*, C1–20.
40. Dollfus, Audouin, “L’observation des étoiles filantes Léonides en ballon de 1898 à 1900”, *l’Astronomie*, **121** (2007), 568–72 and **122** (2008) 20–7.
41. Janssen, Jules, “Note sur les observations des étoiles filantes dites Léonides faites sous la direction de l’observatoire de Meudon”, *CRAS*, **129** (20 November 1899), 788–94.
42. Dollfus, Audouin, “Les premiers ballons-sondes, l’atmosphère et l’astronomie”, unpublished note, September 2006.
43. Janssen, Jules, “Discours prononcé à la séance du 15 septembre 1890 du Congrès international d’aéronautique”, *L’Aéronaute*, **33** (1900), 202–8.

Chapter 6
Janssen and the Sun in Its Majesty: The Eclipse of 1871

> *"Tout d'abord l'auréole. Cercle plus ou moins lumineux qui entoure le disque noir de la lune et qui s'étend souvent à des distances considérables du disque. Ces rayons forment la gloire du Soleil éclipsé".*
>
> *[First of all the aureole. A circle, more or less luminous, which surrounds the dark disk of the Moon, and which frequently extends to considerable distances from the disk. These rays are the glory of the eclipsed Sun.]*
>
> Jules Janssen [1]

The Man Mad About the Sun and his Friend Jules Simon

The next stage for Janssen was the study of the halo, or solar corona. In July 1871, when he compiled his short report for the Minister, Jules Simon, on the British astronomical establishments, he seized the opportunity, at the end of the communication that he sent from London, to talk about the next eclipse:

> Right now, major preparations are being made to travel to India to observe the eclipse of 11 December next, which would undoubtedly recompense us for the failure last year, and which is, moreover, the last phenomenon of this type for quite a long time. The English Committee offered for me to join them, but in ignorance of where I am relative to the intentions of the French government, I have provisionally declined this most flattering and generous offer [2].

Once again, Janssen aroused the patriotic spirit of the man who held the purse strings with regard to expeditions, and this approach was a good one. Thanks to the support of his friends in the Académie des sciences and the Bureau des longitudes, he was again given the task and, as he was to be the only person to represent France, he had at his disposal the 18,000 F that he requested, a tidy sum. But "at the moment of departure and as a result of expenses that he had not expected" [3], he requested an increase in his allowance. The Minister gave in, finally making available 12,000 F, instead of the 4,400 F promised by the expedition department. Jules Simon would,

F. Launay, *The Astronomer Jules Janssen: A Globetrotter of Celestial Physics*, Astrophysics and Space Science Library 380, DOI 10.1007/978-1-4614-0697-6_6,

undoubtedly, be recompensed a 100 times when, 24 years later Janssen provided him with spectacles that gave him back his sight:

> My dear friend, you have been kind enough to send me your lectures. Thank you for them. I am not yet in a fit state to read them, although I manage to write quite easily with your spectacles and two illuminated lamps. I went to the Vaudeville yesterday, in a box, and I saw all the actors. I have been denied such a pleasure for more than a year; I only saw ghosts. So it is you whom I must thank for this extraordinary success. But spectacles for distance viewing are far less interesting to me than spectacles for viewing close at hand, a working tool for me, who does not have any relations on the Moon [4].

This was not the first time that Janssen "introduced a remedy of an optical nature" to correct defective vision: around 1892 he had indeed successfully calculated and had made glasses for the son of a female friend of the family, whose corneas were conical in shape [5].

For the 1871 expedition, the Académie des sciences had given 6,000 F, the Bureau des longitudes 7,600 F, and the "Voyageurs naturalistes" 2,000 F [6] (because Janssen was also responsible for sending a certain number of animals to the Natural History Museum). With the 12,000 F from the Ministry, the sum total of the funds finally allocated to him was thus 27,600 F.

En Route for India

When he left Marseille on 15 October, although he was the only official French scientist to leave, this time Janssen was not the only tourist: Henriette, whose dreams of travel were finally realized, was part of the expedition. She was happy and very proud to be able to play the part that she had considered to be hers ever since she had married, and she described the situation at the beginning of her account of the voyage that she was to publish in 1882 in *Le Tour du Monde:*

> M. Thiers, eager for all of our national glories, wanted France to once again be represented in this scientific contest, and provided all the costs of an expedition to India, which he entrusted to M. Janssen. My being accustomed for a long time to assist my husband in his work, it was decided that I would accompany him in the status of secretary [7].

There is little doubt that this sentence is largely responsible for the status of secretary that is traditionally accorded to Henriette in biographies of Jules. It is true that she would have done more for her husband than her future neighbor at Meudon, the wife of Marcelin Berthelot, whose mortal remains went straight to the Panthéon, accompanying those of her husband. Flammarion did not fail to remark, quite rightly, to Janssen that: "I never saw Madame Berthelot, but it seems to me that your dear and intelligent companions *[Henriette and Antoinette]* have been constantly associated with you, quite otherwise than she with her husband" [8].

Although this situation provides us with an interesting account that would win Henriette many congratulations, it does, on the other hand, deprive us of certain pieces of information that we have become accustomed to find in the couple's correspondence, which was now reduced to an exchange of short notes.

The Choice of Site and Preparations

The aim of this new expedition was therefore "to determine the nature of the corona about which, despite the observations of 1869 and 1870, there still hung many doubts" [9], and the duration of the eclipse, about 2 min, should permit success. Two conditions needed to be present simultaneously to achieve this: it was necessary to find a site where the air would be extremely pure, and to have a very powerful collector of light, to obtain a bright spectrum.

As far as the site was concerned, Janssen had, for a long time, preferred the south of India or the north of the island of Ceylon to New Zealand or Java, because of the more favorable climatic conditions. The first destination was therefore Ceylon, where he landed on 5 November. Once there, he first took care of the Museum's request and he quickly found someone who would undertake to provide him with the desired collection of animals "alive or preserved." Janssen then soon realized that meteorological predictions for the north of the island were not very good, because of morning sea mists, and decided to move to the Malabar coast, to the southwest of the subcontinent. On 20 November, he reached Tellichery (near the French trading post at Mahé), a British port exporting coffee, where he hired two Indians who spoke both French and the local dialects. There was enough time remaining for him to choose his site with some care. Close by, on the Calicut highlands was Nilgiri, the "blue mountains," and we know how Janssen favored high locations. He left his luggage in the central railway station at Coimbatore, and left to explore the mountain range which included summits around 9,000 ft (2,700 m). He learned that in the Nilgiris, the months of December and January were generally fine, especially in the morning. This was ideal, because the eclipse of 12 December was due to take place at 07:30. On the lower slopes, he traveled in an ox-cart. Higher up, because of the infirmity caused by the fall as an infant, he initiated an original system of transport:

> To gain time, with the eclipse approaching, I arranged things so as to use the nights. Sitting in an armchair tied to bamboos, and carried by eight Indians, I could, wrapped in my blankets, sleep during night's trip, which was accomplished at a fast pace, by the light of illuminated torches. Arriving by morning at the site that I wanted to examine, I watched the Sun rise [10].

The Nilgiris were definitely a good site, and Janssen decided to set up in the northwest of the range, on a mountain above Schoolor. He thus avoided the thin curtain of cirrus that always hung above the eastern side. He had all his materiel carried up into "these wild mountains" by a considerable number of porters, but not on their backs: "The crates were fastened with ropes and suspended from bamboos." In a week "the observatory was erected, the instruments mounted, and ready for observation." Janssen explained in his report that "he gives all these details to guide future travelers in conducting such a difficult scientific expedition across a country that is so little civilized" [11]!

As far as the instruments were concerned, Janssen had available those that he could not use at Oran in 1870 and, in particular, the reflecting telescope that he had then had specially constructed. This instrument was very fast because it had an aperture of 38 cm and a very short focal length of 1.42 m. It gave "an image that is 16 times as bright as that of a refractor of the same aperture, and which would have a focal length four times as long," and the instrument did indeed "show in Jupiter details beyond the two well-known, broad equatorial belts" [12]. The associated dispersion system was also very fast. It was a direct-vision spectroscope, built in accordance with the principle of the one in 1862, which consisted of two trains of five very clear prisms, cemented with Canada balsam. In addition, the spectroscope was now fitted with a graduated scale. Moreover, Janssen discovered a simple and clever arrangement to increase efficiency and to allow the same observer (in this case, himself) to follow the whole phenomenon through the finder with one eye, and to carry out the spectroscopic analysis of the region of interest with the other: "all that is necessary is to close one or the other alternately to obtain the image of the region being studied, or the corresponding spectrum" [13]. To study polarization, Janssen also had an "excellent refractor fitted with a bi-quartz, very skillfully constructed by M. Prazmowski" [14]. All the operations were carried out the day before, and the programme for the 2 min was rigidly specified, and on the day itself, Janssen was up by 4 o'clock: "I had rested and, I would add, was full of confidence: I felt that I was ready" [15], he asserted.

The Observation

As was his custom, Janssen gave a very colorful description of the spectacle that the eclipse offered:

> I followed the eclipse in the finder, which was fitted with a very dark glass, which only gave an extremely pale image of the Sun, and leaving my eye with full sensitivity. Totality was approaching. The sky was of a wonderful clarity. [...] Meanwhile the Sun was about to be completely eclipsed; it was currently reduced to a thin, luminous thread, which soon resolved into separate beads.[1] I dropped the dark glass from the refractor, and the corona appeared in all its splendor. Around the Moon there were several fluctuating prominences of a coral pink color, which were seen against the background of a softly luminous halo, white in color, matte, and almost velvety. The edges of this corona were irregular, and quite clearly defined. The overall form was that of a square with curved boundaries, centered on the Sun, and extending outwards from the latter by about half a radius in the lowest portions and nearly twice that at the angles; none of the diagonals was aligned with the solar equator. This corona exhibited a very curious structure which may be used to help to resolve several theoretical points. It was possible to distinguish luminous streaks, which leaving the lunar limb, linked up in the outer portions of the corona; the appearance was that of an ogive or the petal of the flower of a dahlia. This structure was repeated all the way round the Moon and, overall, the corona resembled a luminous, gigantic flower, with the black disk of the Moon occupying the center [16].

Janssen observed that the spectrum of the corona was complex: it consisted of both bright lines – those of hydrogen, and the famous green line (which, in "a remarkable result," appeared to vanish in the spectrum of the prominences) – and dark lines, including the D line of sodium. This observation, and the detection of

[1] These were the famous "Baily's Beads", which were first described by that astronomer in 1836.

a radial polarization (with a maximum effect a few minutes from the limb) indicated the existence, in the coronal spectrum, of reflected sunlight. All his observations enabled him to draw the following conclusions:

> To summarize: it now appears to have been shown, from observations in 1869, 1870, 1871:
>
> That the phenomenon of the corona at total eclipses is caused by a gaseous envelope belonging to the Sun;
> That this envelope is self-luminous, at least in the portions close to the Sun;
> That it has an extremely low density and a temperature much lower than that of the chromosphere[2];
> That the gas hydrogen is the principal element;
> That this gaseous envelope is definitely not in a static state [...];
> Given that this layer forms an envelope that is very distinct from the chromosphere, there is cause to give it a name. I propose to call it the *coronal atmosphere* [17].

This time, Janssen was better at communicating. He sent several telegrams, not only to the secretariat of the Institut, but also to the Minister, and he did not slip French words into his sentences written in English: he either wrote just in English, or else he sent his text in both languages.

He also sent his results to Warren De La Rue, the Royal Society, the Royal Astronomical Society, and to the British Association "which had been so benevolent towards him" [18].

However, it must be admitted that his subsequent reports do not give as much information as one might wish. Janssen described very precisely the choice of the observing site, which was excellent and which was to have a future [19], his perfectly chosen instruments, and the beauty of the corona. He drew fairly precise conclusions, but did not linger over them. He provided an illustration, but did not indicate its nature: is it a photograph, or a drawing? Admittedly, he did not say anywhere that he took photographs, and we know that he was good at drawing. It is only thanks to a British publication that we have all the details that we might want. There, in the *Memoirs* of the Royal Astronomical Society [20], the reporter who reproduced the original French text requested written details. This is how we learnt from comments in English that the representation of the eclipse is a wood engraving from one of Janssen's drawings; the orientation is given, as is the position of the spectroscope slit during the observations, which is shown on a diagram. The pattern of the observed lines is also added (Fig. 6.1). It is thus easier to appreciate the efficiency of Janssen, who did everything: overall visual observation with one eye, spectroscopic observation with the other, and who took the time to make a drawing, to admire, to dream, to record everything in his head, and who still had a few seconds to make polarimetric observations! Reading the British reports, we realize at the same time that Colonel Tennant, assisted by Lieutenant John Herschel, had a lot of trouble and finally did not observe very much, at a site that was far worse chosen, even though he was a native of India, where he had always lived.

---

[2] We now know that it is nothing of the sort, and quite the contrary, because the temperature of the corona is greater than a million degrees [°C], whereas that of the chromosphere is only about 6000 degrees [°C].

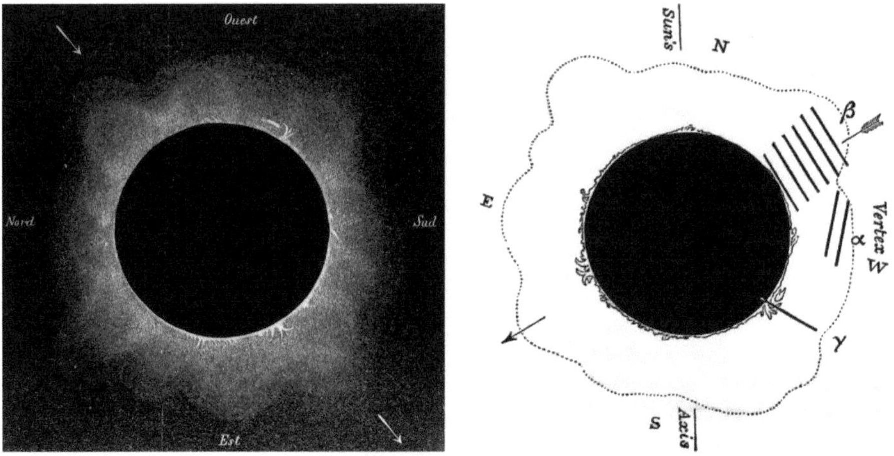

Fig. 6.1 *Left: a drawing of the eclipse of 1871 made by Janssen.* Annales de Chimie et de Physique, *1873. Right: a drawing by Janssen of the eclipse of 1871, as annotated by the British.* Memoirs of the RAS, *1879.* © Paris Observatory Library

And it was the British who paid tribute to Janssen, who did indeed see several absorption lines in what we now call the "F corona" [21], because his spectrum was characterized by the same lines as direct sunlight (the Fraunhofer lines), when Herschel saw just one [22].

Although usually always early with a technique, curiously Janssen did not carry out any scientific photography. The British superiority thus rests with the fact that they obtained images, and they are particularly spectacular, but, like Janssen, they were only able to publish engravings [23].

Lockyer, also on an expedition to India, had said in September how pleased he would have been for Janssen to observe the eclipse with him as he had been invited to do, but "after all" he concluded, "it is better that he should represent France, which he will do most nobly" [24]. Stopping in Paris on his return journey, he informed Dumas that he had the pleasure of seeing Janssen in India before he left [25], which proves how cordial relations were between the two "who revolutionized spectral analysis," which was how Wilfrid de Fonvielle described the two friends [26] (Fig. 6.2).

Henriette and the Todas

Janssen thus returned without any objective image of the astronomical phenomenon. On the other hand, and in addition to the animals, he brought back for the Museum photographs of the inhabitants of the Nilgiris taken by others [27]. During slack periods, and while his wife, who had joined him at the observation site after the eclipse, observed the customs of the natives and gathered information, he took cranial measurements and filled his notebooks with drawings [28]. The one that shows a "Toda type" was reproduced in facsimile (Fig. 6.3) in the article "Les Todas"

M. JANSSEN M. LOCKYER

Les réformateurs de l'analyse spectrale.

Fig. 6.2 *The men who revolutionized spectral analysis.* Le Monde Illustré, 26 September 1874. © Author's collection

published by Henriette in the *Tour du Monde*, which was run by the senator Édouard Charton, while others would be given the same treatment as the British scientific photographs and were reproduced by gravure.

The Janssens' relationships with the Toda tribe were not made until after the eclipse. They were that much better given that the members of the scientific expedition did not deny that with their instruments they had seen the struggle of the great dragon to devour the Sun, to which the Todas attributed the Sun's disappearance during total eclipses.

Of course, Mme Janssen had no pretensions to make her narrative of 1882 into a scientific publication. She did not hide the fact that she had profited from the acquaintances of her English host, Breeks, and that she had taken some details from the book by the German missionary Metz, which had appeared in 1864, but her account, which makes pleasant reading, could not fail to captivate her readers. Among those were distinguished names such as that of the cousin, Eugène Labiche: "Please thank Madame Janssen for the copy of the *tour du monde* that she sent us. Her work on the *todas* is very curious, and I must add that it is written with a simplicity and a sincerity that are full of charm." [29], or again, that of Victor Duruy:

> I had already read le Voyage aux Nilgherries and, at the Institut, I sought out M. Janssen to congratulate him on this other fame that had been added to his own. You have been kind enough to send me a copy: I am grateful to you for this gift and souvenir.
>
> Please accept, Madame, my respectful congratulations.
>
> The devoted servant of the married couple who serve Letters and Science so well [30].

Fig. 6.3 *A* "Toda type". *A drawing by Jules Janssen.* Le Tour du Monde, *1882.* © Author's collection

And, in addition to the respect of great men, the article brought Henriette 15 centimes a line [31]!

References

1. Janssen, Jules, "Conférence sur l'éclipse totale du 18 août 1868", in *Lectures académiques discours*, 299–329, p. 312.
2. Letter from Jules Janssen to Jules Simon, 22 February 1871, AN F^{17} 2977.
3. Note for the Minister of Education, 05 October 1871, AN F^{17} 2977
4. Letter from Jules Simon to Jules Janssen, *ca* 1894, BIF Ms 4137-039.
5. Janssen, Jules, "Sur les dispositions optiques propres à remédier aux troubles visuels dans les cas de kératocône", *CRAS*, **134** (5/20 May 1902), 1121–3.
6. Ministère de l'Instruction publique, note du 12 octobre 1871, AN F^{17} 2977.
7. Janssen, Mme, "Souvenirs d'un voyage aux Nilgherries, les Todas", *Le Tour du Monde*, **XLIV** (1882) 1137e livraison, 241–9.
8. Letter from Camille Flammarion to Jules Janssen, 21 June 1907, BIF Ms 4135-179.
9. Janssen, Jules, "Résultats de mes observations dans l'Inde, sur l'éclipse du 12 décembre 1871" in *Report of the 42nd Meeting of the BAAS, Brighton, 1872*, 34–5.
10. Janssen, Jules, "Rapport sur l'éclipse du 12 décembre 1871, observée à Schoolor (Indoustan)", *Annales de Chimie et de Physique*, **28** (1873), 474–99.

11. *Ibid.*
12. *Ibid.*
13. *Ibid.*
14. *Ibid.*
15. *Ibid.*
16. Janssen, Jules, "Rapport sur l'éclipse du 12 décembre 1871, observée à Schoolor (Indoustan)", *Annales de Chimie et de Physique*, **28** (1873), 474–99.
17. Janssen, Jules, "Résultats de mes observations dans l'Inde, sur l'éclipse du 12 décembre 1871" in *Report of the 42nd Meeting of the BAAS, Brighton, 1872*, 34–5.
18. Letter from Jules Janssen to Warren De La Rue, 19 December 1871, Cambridge University Library (CUL), RGO Archives, RGO 6/132.
19. Launay, Françoise and Hingley, Peter D., "Astronomy", in *Encyclopedia of the Nilgiri Hills*, Paul Hockings, ed., New Delhi, Manohar Books, p. 84–90, to be published.
20. Janssen, Jules, "Rapport à l'Académie relatif à l'observation de l'éclipse du 12 Décembre", *Memoirs of the RAS*, **41** (1879), 450–4.
21. Bappu, M.K.V., *Astronomy in India during the period 1787–1947,* Ramakrishna Mission Lectures, 1974.
22. Tennant, J.F., "Report on Observations of the Total Eclipse of the Sun on December 11–12, 1871, made by order of the Government of India, at Dodabetta, near Ootacamund", *Memoirs of the RAS*, **42** (1873–5), 9.
23. *Memoirs of the RAS*, **41**, (1879).
24. Letter from Norman Lockyer to Jean-Baptiste Dumas, 11 September 1871, AAS, Lockyer file.
25. Letter from Norman Lockyer to Jean-Baptiste Dumas, 1872, AAS, Lockyer file.
26. De Fonvielle, Wilfrid, "Les deux réformateurs de l'analyse spectrale", *Le Monde Illustré*, No. 911, 26 September 1874, 203 and 208.
27. Mahias, Marie-Claude, "Le Soleil noir des Nilgiri, l'astronomie, la photographie et l'anthropologie physique en Inde du sud", *Gradhiva*, **24** (1998), 45–56.
28. Launay, Françoise, "Autour de l'éclipse totale de soleil visible in Inde le 12 décembre 1871: le voyage du couple Janssen aux Neelgherries", *l'Astronomie*, **111** (1997), 111–5.
29. Letter from Eugène Labiche to Jules Janssen, 23 February 1883, BIF Ms 4136-429.
30. Letter from Victor Duruy to Henriette Janssen, 28 October 1882, BIF Ms 4135-95.
31. Letter from Édouard Charton to Henriette Janssen, 28 October 1882, BIF Ms 4135-12.

Chapter 7
Janssen and the Cinema: The Transit of Venus of 1874 and the Revolver Photographique

"Le revolver photographique donne au contraire [du phénakisticope (sic)] l'analyse d'un phénomène en reproduisant la série de ses aspects élémentaires".

[The photographic revolver provides, on the contrary [to the phenakistiscope the analysis of a phenomenon by reproducing a series of its individual elements.]

Jules Janssen [1]

Although later he would repeat relentlessly that "celestial photography and spectral analysis today form the two branches of physical Astronomy" [2] Janssen, as we have seen, did not, however, use photography in 1871. And suddenly, faced with the problem presented to astronomers when they were preparing to observe the transit of Venus in December 1874, he added another string to his bow and straightaway perfected nothing less than the prototype camera for taking cinematograph frames!

With this invention, which was of the greatest significance, and which he baptized the "revolver photographique" *["photographic revolver"]*, photography was brilliantly introduced into his professional life, and soon acquired fame among the British, where the name of the instrument quite simply became the "Janssen" [3].

The Problem to be Solved During the Transit of Venus

The British astronomer Edmond Halley had suggested many years earlier that if observations of a transit of Venus in front of the Sun could be carried out from two widely separated locations on the Earth, and the times of the corresponding contacts measured accurately, it would be possible, calculating by trigonometry, to improve the determination of the average distance between the centers of the Sun and the Earth. This value, now called the "astronomical unit" is fundamental, because Kepler's laws do not allow astronomers to determine the average radius of the orbits of planets

F. Launay, *The Astronomer Jules Janssen: A Globetrotter of Celestial Physics*, Astrophysics and Space Science Library 380, DOI 10.1007/978-1-4614-0697-6_7,

except as being relative to one another. To put an accurate figure to the distances that separate them, it is thus necessary to know the true average distance between the Sun and one of them.

But a transit of Venus across the disk of the Sun is one of the rarest phenomena that occur in astronomy. Indeed, although two consecutive transits occur with an interval of 8 years, each pair is separated from the next one by 105.5 and 121.5 years alternately.

In the eighteenth century, a lot of energy had therefore been expended, by both institutional and other observatories, to travel to observe the transits of 1761 and 1769, and the expeditions of Captain Cook, the voyage of Abbé Chappe d'Auteroche, and the peregrinations of Le Gentil that were carried out and succeeded (or failed ...) have become famous [4]. All the efforts were, unfortunately, in vain, because the drawings made from visual observations, and the measurements of the times of contact were insufficiently accurate because of the infamous "black drop" effect [5], which made the limb of the planet appear to be linked to that of the Sun by a dark thread, which made any accurate determination of contact impossible. The difference between the duration of the transit across the solar disk from one site on Earth to another, even well separated, is very small, and the uncertainty of the measurements, which was about 1.5% of the value (about 8.8″) of the solar parallax (the angle subtended by the Earth's radius as seen from the Sun: from the Sun, the Earth appears similar to a coin, 2.5 cm in diameter, seen from a distance of 300 m) [6]. The error was thus ±0.13″, and remained too great.

Nineteenth-Century Aspirations

More than one hundred years later, techniques had made such progress with the development of photography on the one hand, and that of the electric telegraph (which allowed clocks to be synchronized with those of a nearby observatory) on the other, that astronomers were certain of doing better. They hoped for an accuracy of ±0.01″ for the parallax. That is why the expeditions sent to observe the transits of 8/9 December 1874 and 6 December 1882 probably gave rise to the greatest international effort ever mounted up to that time, to observe a "simple" astronomical event. Everywhere, special commissions produced series of reports aimed at determining the most favorable observation sites, and at agreeing on the best tools to be used.

In 1874, there were no less than 62 expeditions, representing some ten countries, which spread out over some 80 observation sites. All the major maritime nations were on board, and ships carrying astronomers and equipment left for sites that were scattered as widely as possible in both the northern and southern hemispheres. In France, the government asked the Académie des sciences to set up a committee to co-ordinate operations. All the members involved were drawn from the astronomy section, or that of geography and navigation. Among the astronomers was Charles Delaunay, the chairman of the committee, who, after his tragic death in 1872 (he died in a drowning accident), was replaced by Faye; Le Verrier and Janssen, after he was elected to the Académie des sciences, on 10 February 1873 (Fig. 7.1).

Fig. 7.1 *Janssen, new member of the Académie des sciences.* L'Illustration, *15 February 1873.* © Author's collection

At that time, Dumas became head of the committee, replacing Faye, who had other activities. The committee finally decided that six French expeditions should be sent to, respectively, the southern islands of Campbell and Saint-Paul, to Noumea, to Peking, to Saigon, and finally, to Japan (with Janssen in charge). As, among the 50-odd Frenchmen who were leaving, he was the sole member of the Académie des sciences, Janssen also played the part of the official representative of the institution [7].

Janssen and His Revolver

This was by no means a mean feat because Janssen did not cease distancing himself from his colleagues by proposing solutions different from those adopted by the committee...

Photography was, of course, soon chosen as a means of resolving the difficult problem of accurately determining the contacts. Instead of having an observer giving a mark, an estimate of greater or lesser accuracy, at the instant of contact, and a helper recording the corresponding time, it would suffice to record the exact time that each image was exposed, and which could be examined later. To facilitate reduction of the results, the committee therefore decided that each French expedition would be provided with an identical "national apparatus." But this photographic equatorial was not at all to Janssen's satisfaction, because it gave images that were much too small for his taste: the Sun measured just 3.5 cm in diameter. As a result, Janssen had a special instrument built for his expedition "with the extremely thorough assistance of M. Prazmowski, whose skill and knowledge of optics are well-known"

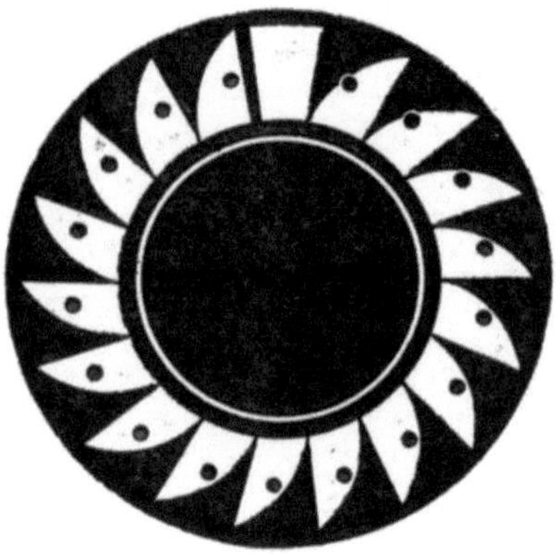

Fig. 7.2 *Diagram of the principle of the revolver.* L'Illustration, *30 May 1896.* © Author's collection

and with "the skilled photographic operator M. Arents" [8]. He was thus able, like the members of the foreign expeditions, to photograph large images that he obtained with a refractor of 10 cm aperture and a focal length of 2 m, and which he enlarged to 11 or 12 cm in diameter "using an eyepiece." On 22 June 1874, he displayed photographs obtained with this instrument, of which he was particularly happy.

Could he not, however, do even better? Could he not, for example, design a mechanical arrangement capable of automatically *recording*, on the same photographic material, at each of the four contacts, a sequence of images taken at very short and regular intervals? There would then only be the need to record the time of the first exposure, and the photographic plate, the equipment's mechanism and, after the expedition's return, a microscope would do the rest. "One would thus substitute photographs, which could be examined later at one's leisure to deduce the true instant of contact, for fleeting appearances, which an observer, of necessity nervous, would have to interpret instantaneously" [9] (Fig. 7.2).

The idea of combining chronometry and photography had been on his inventor's mind for some time, and on 15 February 1873, just 5 days after his election to the Académie des sciences, Janssen presented the committee with a method, which would eventually solve the difficult problem of determining the exact times of the planet's contacts with the solar disk by a simple, objective, and durable means. By using an annular plate, Janssen planned to record 180 images on a disk, at the rate of one per second, with the appropriate exposure time being obtained by shutter having a variable slit. "It would thus be possible to start the photographs one-and-a-half minutes before the assumed instant of first contact [...] and do the same for all four" [10].

The idea of recording several images on the same plate was certainly not new, but none of the instruments constructed on this principle, such as Thompson's "revolver" built by Briois [11] and the "pistolgraph" by Skaife [12], which Janssen quite obviously did not know about, allowed sequential image capture.

After his account of the principle of the "contact apparatus," Janssen modified his plans to have the instrument built. The disk would no longer record 180 images, but just 48, and the rotating plate would no longer be driven by electricity, but by a spring. The first form of the instrument by the maker Eugène Deschiens, although "carefully produced," did not satisfy Janssen because of vibration created by the fact that the disk with the interrupting slit did not have a constant motion. Janssen obviously did not obtain the modifications that he requested from Deschiens, whose workshop he frequently visited. (He probably had some brush with him, just as he had had with Hofmann 10 years earlier.) So he made the instrument makers Redier, father and son, responsible for constructing the new version of the instrument [13]. While the plate carrying the annular daguerreotype rotated by 1/48 of a turn, then remained stationary during the exposure (what Janssen called an "alternating movement" [14]), thanks to a Maltese-cross mechanism, the disk that carried the 12 radial interrupting slits (regularly spaced, and adjustable in width) rotated continuously at four times the rate. So when the photographic plate had finished a complete turn (in 72 s), the interrupting disk had made four rotations. It was with this second version of the instrument, thenceforward baptized the "photographic revolver," that Janssen was able to carry out tests of photographs of an artificial transit, a specimen of which he displayed on 6 July 1874 that "proves that the images may be obtained with great clarity" [15]. Three examples of these daguerreotypes have been preserved. One of these plates is in the collection of the Société française de photographie. The second is at the Observatoire de Paris, together with an example of the revolver. The third plate, bequeathed to the CNAM[1] by Antoinette Janssen with another example of the Redier revolver, is the one that Joseph Leclerc, the founder of the film archive of the Société astronomique de France, used to reconstruct the motion of Venus in one of his films [16]. An example of the Deschiens' version of the revolver is also preserved in the collection of the French Film Archive [*Cinémathèque française*] [17].

The photographic revolver was extremely well described in 1875 by Flammarion in *La Nature* (Fig. 7.3). A peerless popularizer, Flammarion wanted to be understood, and to be understood he pressed the scientists he was questioning to give him all the necessary explanations. It is thanks to him that we understand the nature of the "alternating movement", and it is also thanks to him that we learn the nature of the plates that were employed, namely daguerreotypes, and the justification for their use, which however, left the reader hungry for more: "M. Janssen preferred the Daguerre process to photographs on paper, because of the greater clarity of the images on the silvered plate" [18] No mention of the advantages relative to wet collodion plates, which would, of course, be used for the usual photographs, nor relative to albumen plates. Flammarion did also mention the revolver's Maltese-cross mechanism, which Janssen did not name explicitly, and in addition he specified the key factor that "a significant accessory enables the exact time of each exposure to be recorded".

As for him, always more preoccupied with the future than with the past, Janssen spent far more time describing his plans than in giving an account of the details of

[1] The Conservatoire national des arts et métiers [the Museum of Technology].

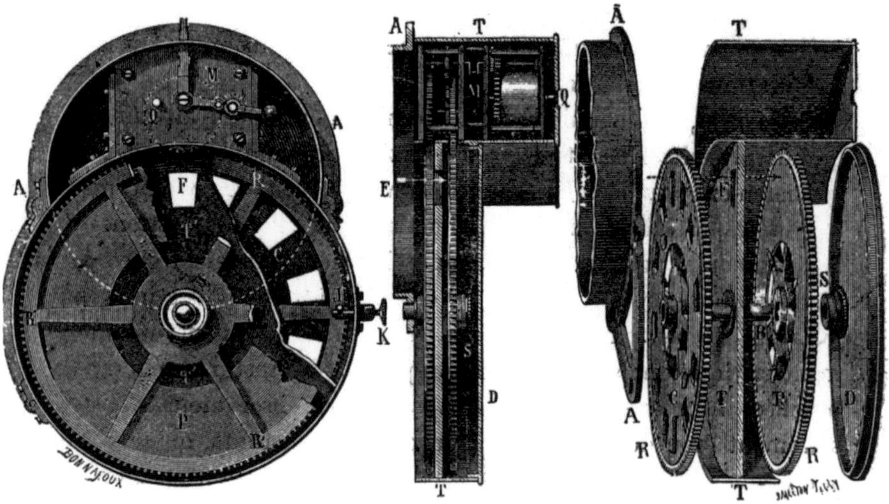

Fig. 7.3 *Exploded diagram of the revolver.* La Nature, *08 May 1875.* © Author's collection

what he had achieved. If he eventually decided in April 1876 to display his revolver to the Société française de photographie, which he had joined a week earlier, it was for a very specific reason: a committee had been appointed on 16 March, the month before, to examine his accounts, because he had spent so much in having his additional instruments built! [19] (Fig. 7.4).

Yet, at the same time, it would have been possible to read, in the *Journal officiel de la République française*, dated 8 June 1875, this sentence, written by Dumas, the chairman of the committee, and which would bring both pride and sadness to a French reader: "M. Janssen has, in addition, been able to put the ingenious apparatus that he has invented – and which the English astronomers have named after him – to use, and which the English expeditions have all been able to make the most of, and which lack of time as well as finance have prevented us from providing for all our own expeditions"!

The British and Their "Janssens"

In fact, as soon as he knew about Janssen's project in the spring of 1873, the Astronomer Royal, Sir George Biddell Airy, took a close interest. On 24 April, he asked the famous London instrument-maker Dallmeyer to consider the possibility of adapting such equipment for his photoheliographs [20]. He was, of course, in touch with De La Rue, who asked his friend Janssen to send him drawings of the apparatus. In June, Janssen promised to send them [21], but then forgot to do so.

Fig. 7.4 *Janssen's revolver. A contemporary photograph.* © Paris Observatory library

When returning from the September BAAS meeting in Bradford, he stopped in London, and a meeting with Airy and De La Rue took place at Greenwich on 6 October:

> I have given Sir G. Airy and De La Rue, details of the construction of my apparatus, and how it may be adapted for their instruments. They have already made some preliminary trials, which have been successful. I also have an exact list of their observing sites and of their equipment – the addresses of the best English makers of the instruments that we may require [22].

We know how scathing Sir George could be, and he gives a particularly significant example in the letter that he wrote several weeks later to De La Rue:

> If we are to use Janssen's method at all, we must decide on our plans speedily. Janssen has not the least idea of mechanics, and having got his principle, I throw him overboard. Now you understand both photography and mechanics better than I. I will not throw you overboard, and should be much pleased to know your plans. I should prefer (generally) independent manual motion of each moveable part in order to avoid complexity of apparatus [23].

At the end of January 1874, Airy ordered a prototype from Dallmeyer, while De La Rue, for his part, undertook the design and construction of two different, automatically operated, instruments [24]. The first, destined for India, was capable of recording 60 images per plate, while the second, destined for Melbourne, was smaller, and recorded 20. Airy, who preferred Dallmeyer's 50-image model with manual rotation, ordered four more pieces of equipment. The last was delivered toward the end of May 1874, even before Janssen had the use of Redier's apparatus. Airy was perfectly satisfied with all his "Janssens", the use of which nevertheless required no less than four people per instrument …

The Trip to Japan

In Paris, Janssen assembled the team [25] that would accompany him on the expedition, and all the instrumental trials were carried out at a yard on the boulevard Ornano, which he had already been using for several years to store the instruments for his eclipse expeditions. At the end of July 1874, a photograph recording the members of the expedition was taken, probably by one of the journalists from *L'Illustration* or *L'Univers illustré*, which were copiously covering the preparations for the French expeditions [26]. In the photograph, Janssen, seated in the center, presides (Fig. 7.5). One of the observers of the 1868 eclipse, with whom he had already traveled, the

Fig. 7.5 *Janssen and the team for Japan. A photograph taken in Paris.* © Anne Guigan-Léauté Collection

astronomer Félix Tisserand, who became Director of Toulouse Observatory in 1873, is seated to his right. Behind them are the Japanese Chimizou (as Janssen spelled it), alias Makoto Shimizu, who entered the École centrale des arts et manufactures de Paris in 1873, but who, after the "transit" would remain in Japan, where he distinguished himself as a pioneer in the local match industry; the Brazilian astronomer "D'Almeida", alias Francisco Antônio de Almeida, an interim lecturer at the Rio Polytechnic, whose addition to the expedition had been accepted by the Académie des sciences at the request of the emperor Dom Pedro II; and the helmsmen Mercier and Michaut, recognizable by their sailor's uniforms. The two other persons are probably the photographer Pierre Marie Arents and Lieutenant Jean Picard, a graduate of the École polytechnique. Missing from the roll call are Sub-Lieutenant Eugène Delacroix, who, benefiting from a late recommendation by a deputy, who was a friend of his father's, joined the group only at the last minute at the end of August; and the mechanic Vacher.

Once again Henriette was on the trip, and it was she who wrote the letter to the Académie des sciences from Hong Kong, describing the two dreadful storms, 3 days apart, from which the travelers had suffered. It was the second that was particularly impressive and deadly. During the night of the typhoon when "Janssen was overwhelmed by sea-sickness" [27], the ship could only remain off Hong Kong, its sole hope being that the anchor-chain would not break – which, luckily, was the case. The next day, however, Henriette described the distressing sight that the expedition's members experienced:

> The town was still under water; the sea, which was extremely rough, bore all sorts of wrecks towards us; just spars and hulls of boats appeared where, the day before, we had seen whole, smart-looking vessels. And for two days they have continued to collect floating bodies around us. More than fifteen hundred Chinese have disappeared with their sampans, which are small boats where whole families live. A Spanish ship that arrived this morning had lost ninety passengers and its crew. A dozen ships are reckoned to have been lost [28].

At Hong Kong, the group embarked on another ship, sailing to Yokohama, where it had been decided to locate the administrative base for Janssen's expedition. As always, Janssen did not know where he would site his instruments, but we know that he was in the habit of arriving sufficiently early to study the climate on site with local people, and to choose the most favorable place. Initially he preferred Kobe, but upon investigation, it transpired that predictions were better for Nagasaki. So it was there that he settled, on a hill "well clear of the fumes from the town" [29]. The only difficulty was transporting the 250 crates of instruments and various materials (including portable observation cabins), but it did not take him long to find the necessary 500 porters. He was confident, and wrote to Dumas:

> With this mailing I am sending news to the Académie.
>
> You will see there that we are at Nagasaki, actively working on our installation, which will be finished in a few days. Our voyage has been rather checkered and we have not spared ourselves in trying to turn everything to our advantage.
>
> At present, all that remains is to pray to Him who commands the waves and, I would add, the clouds, to favor us on 9 December. If so, we will easily pardon him for our odyssey between China and here.

> This French expedition will probably have an excellent effect here. I have not neglected anything that would tend to such an end [30].

Precisely to increase the chances of success on the day of the transit, he did decide to divide his team and to send Delacroix and Shimizu back to Kobe.

Observations and Results

At Nagasaki, it was Tisserand (who had been first in his year at the École normale supérieure, and been well trained in geodetic measurements during the 7 years that he had spent at the Paris Observatory), who was put in charge of the determination of the longitude and latitude, where a high degree of accuracy was indispensable, as well as being responsible for the clocks and chronometers. He also observed the contacts, as did Janssen, who was unfortunately unable to carry out the spectroscopic observations that he had intended, for lack of an instrument. A gust of wind had broken one of his objectives. While Picard was in charge of the Committee's photographic equipment, Arents took care of the Prazmowski's special photographic refractor, and Almeida was at the revolver's controls. Everyone conscientiously carried out their appointed tasks, even if the atmosphere within the group was not always completely cordial. The request that Janssen made to the Minister on his return to France for "a severe official warning" to be given to Lieutenant Picard is proof of that. The reason was that Picard had, during the expedition, "assumed complete independence" and had "on several occasions, been seriously disrespectful of him [*Janssen*]" [31].

At Kobe, where the weather finally turned out to be better, Delacroix observed the four contacts, only the last being uncertain, while Shimizu took photographs. Overall, the two teams succeeded in taking some 80 photographs of the transit, of which 60 daguerreotypes are today preserved at the CNAM, and whose quality is not as "passable" as Tisserand had reported to his friend Léauté [32]. As far as the revolver is concerned, several tens of "Janssen" plates were obtained around the world. Alas! of these, no original example has yet been located, nor the only annular daguerreotype that Almeida succeeded in recording at Nagasaki, and regarding which Janssen reported that "the weather was then slightly hazy, such that the images are weak, but clearly visible" [33].

After measurement of all the plates and analysis of all the measurements, the French results proved to be no better than those of other observers [34]. Astronomers were disappointed, because although the error bar obtained for the solar parallax was less than the eighteenth-century one, it was still $\pm 0.06''$, or six times greater than they had hoped.

However, undoubtedly more interested in physics than in celestial mechanics, Janssen was particularly proud of being able to state that, on the one hand, he had been able to observe the eclipse of the solar corona by Venus, before first contact, by using a blue-violet filter [35] and, on the other, that "he observed facts which prove the existence of an atmosphere to Venus" [36].

He also set a score on being able to leave material traces of his expedition in Japan. At Nagasaki, he had a pyramid built on the hill at Kompira-Yama, and the governor of the province had a column erected at Kobe. Despite the dreadful suffering that the two localities have undergone in the twentieth century, both monuments are still standing as witness to the Janssen expedition's visit to those sites [37].

The Eclipse of 8 April 1875 in Siam

After the transit of Venus, while his companions started back to France, Janssen bought several souvenirs (in particular, he brought back a cloisonné jar for his friend Henner, in which he could keep his tobacco [38]), and continued his trip, accompanied only by Henriette. How could he not go to Burma or Siam, where an eclipse of the Sun would be visible a few months later? When he arrived at Singapore, he learned that, by agreement with Dumas, Lockyer had made the necessary arrangements for his baggage to be carried to Burma at a reduced rate, but he waited in vain for news from the British astronomical committee. So he decided to leave for Bangkok, to which a "commercial" ship – as he termed it – could carry him [39].

There, he was able to confirm the existence of the corona, and the fact that, in it, the green "1474" line was far more intense than in prominences. He also wanted to make measurements of the extent of the corona, but because he – who had just been experimenting with sequential photography – no longer had either a photographic equatorial, or a photographer, he asked his English colleague Dr Schuster to take photographs of the corona with exposure times increasing in a geometrical progression of 1, 2, 4, and 8 s. He was thus able to demonstrate that the longer the exposure time, the greater the height of the corona that was recorded [40].

The following August, Janssen attended the Bristol meeting *[of the BAAS]*, where he spoke about the transit of Venus, the revolver, the eclipse, and also about mirages seen at sea. It was there that he learned, through Williamson, that he had been nominated as a Foreign Fellow of the Royal Society of London. The following year, at Glasgow, "the expedition to Japan was the only one of the French expeditions to be mentioned and to which particular attention was given" [41].

The Revolver's Future

Although the results obtained by classical photography and by the revolvers used in 1874 for the transit of Venus were hardly conclusive, Janssen had conceived and employed the first piece of cinematographic equipment which, from the outset, he intended for many other subsequent uses: recording of the successive phases of total or partial eclipses; automatic monitoring of the solar disk; the study of the motion of solar granulation [42]; the systematic examination of the circumsolar region for the transits of hypothetical intra-mercurian planets across the Sun [43]; or even the study of physiological mechanics: the locomotion, flight, and various movements of animals [44]. From then on, the way to "animated photography" was wide open to Muybridge, Edison, Marey [45]; and the Lumière brothers!

However, Janssen remained rather jealous of his invention. He obviously knew Jules Marey very well, given that he came into contact with him both at the Académie des sciences and at the Société française de photographie, and each would be President of both one and the other, but the two obviously did not associate very much with one another. When Marey announced to the Académie des sciences, in a letter read on 13 March 1882, that he had just obtained "by means of instantaneous photography, a complete analysis of various forms of locomotion, including the flight of birds", thanks to the instrument that he had finally succeeded in constructing "in the form of a hunting rifle" [46], Janssen stated during the session that he was delighted by it. Marey clearly set out the origin of his instrument and the difficulties that he had to surmount:

> The analysis of the movement of flight demands a series of positions with the knowledge of the exact instant at which each is produced. A method that our fellow member M. Janssen devised at the time of the transit of Venus across the Sun, to follow the phases of the relative motion of these two bodies, seemed to me to fulfill, in principle, the desired conditions. But I found many difficulties in the execution of this, because of the extreme speed at which the pieces of the photographic revolver that I required, needed to move.

However, Janssen could not refrain from exaggerating: "Recently our colleague was kind enough to write to me asking for details of the arrangement of the photographic revolver, and I hastened to reply" [47]. Janssen seems to have been cut to the quick, and 3 weeks later he went a bit overboard and put down a marker with a communication on the principle for a new revolver that would allow the study of… the movement of the wings of insects! To record very rapid movements, he suggested giving the plate a continuous motion, and said that he had already made a successful trial, finishing by photographing the solar granulation on a plate moving at 20 cm per second [48].

In any case, the "animated head" on the circular plate taken with Marey's equipment with a rotating photographic plate, was rather an odd "insect" because it showed a man in a turban, smoking a cigarette (Fig. 7.6). That photograph appears in two of the magnificent albums by Marey that are held by the Collège de France in Paris, and one of the two examples includes Marey's manuscript annotation: "1 January 1884. Successive circular images on a plate rotating in uniform motion (Janssen subject). slow rotation" [49]. The two men did not seem to have been annoyed with one another on that day, if the date is correct, and if Janssen was really involved, because it was customary for Janssen to remain at home on New Year's Day to receive visitors.

To study the duration of a spark, or to determine, in the laboratory, the time that had elapsed between two sparks, in 1888 Janssen suggested a method derived from that of the revolver: thanks to two identical objectives he recorded simultaneously two images of the luminous event on two disks carrying sensitive film, one fixed, and the other rotating rapidly. After initial trials had given interesting results about the sparks produced by a Ruhmkorff coil, he considered using the equipment to

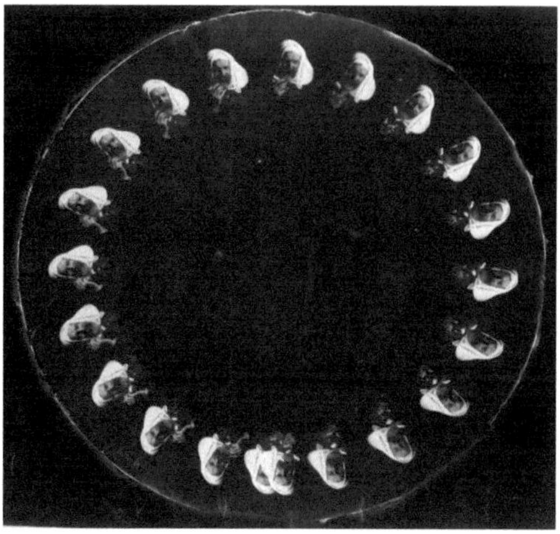

Fig. 7.6 *Marey's circular plate, 01-01-1884. "Sujet Janssen"*. © Collège de France Archives

study lightning [50]. Janssen did not give a name to the apparatus which would have been constructed by Chevallier [51], but there is no doubt that this was what we now, in laboratories, term a "revolving-drum camera".

References

1. Janssen, Jules, "Présentation du revolver photographique et épreuves obtenues avec cet instrument", *Bulletin de la Société française de photographie*, **22** (April 1876), 100–6.
2. Janssen, Jules, "Revue annuelle d'astronomie", *Revue générale des science pures et appliquées*, 30 May 1896, 485–8.
3. Launay, Françoise and Hingley, Peter D., "Jules Janssen's 'revolver photographique' and its British derivative, the 'Janssen Slide'.", *Journal for the History of Astronomy (JHA)*, **36** (2005), 57–76.
4. Dumont, Simone, "Les expéditions du XVIIIe siècle pour les passages de Vénus", *l'Astronomie*, **118** (2004), 290–5. Marlot, Christophe, *Les Passages de Vénus*, Paris, Vuibert, 2004. Toulmonde, Michel, *Les rendez-vous de Vénus*, CD-ROM, Observatoire de Paris, 2004 ; Sheehan, William and Westfall, John, *The Transits of Venus*, Amherst, NY, Prometheus Books, 2004.
5. Horn d'Arturo, Guido, "Il fenomeno della 'groccia nera' e l'astigmatismo", *Pubblicazioni dell'Oservatorio astronomico della R. Università di Bologna*, **1** No.3 (1922), 25–57 ; Pasachoff, Jay M., Schneider, Glenn and Golub, Leon, "The black-drop effect explained," in *Transits of Venus: New Views of the Solar System and Galaxy*, IAU Colloquium No. 196 (U.K., 2004), D. W. Kurtz and G. E. Bromage, eds., 2005, pp. 242–253.
6. Toulmonde, Michel, "La parallaxe du Soleil", *l'Astronomie*, **118** (2004), 274–89.
7. Dumas, Jean-Baptiste, "Rapport sur l'état des préparatifs pour les expéditions chargées par l'Académie d'aller observer le passage de Vénus sur le Soleil, le 9 décembre 1874", *CRAS*, **78** (29 June 1874), 1796–1806.
8. Janssen, Jules, "Présentation de quelques spécimens de photographies solaires obtenues avec un appareil construit pour la mission du Japon", *CRAS*, **78** (22 June 1874), 1730–1.
9. Janssen, Jules, "Présentation du revolver photographique et épreuves avec cet instrument", *Bulletin de la Société française de photographie*, **22** (April 1876), 100–6.

10. Janssen, Jules, "Passage de Vénus; Méthode pour obtenir photographiquement l'instant des contacts, avec les circonstances physiques qu'il présentent", *CRAS*, **76** (17 March 1873), 677–9.
11. Harding, Colin, "Images on disc", *Photographica World*, No.62, September 1992, 7–12.
12. Hingley, Peter D., "Thomas Skaife pistolgraphs", *Astronomy and Geophysics*, **40** (1999), 5.7.
13. Janssen, Jules, "Présentation d'un spécimen de photographies d'un passage artificiel de Vénus obtenu avec le revolver photographique", *CRAS*, **79** (06 July 1874), 6–7.
14. Janssen, Jules, "Présentation du revolver photographique et épreuves obtenues avec cet instrument", *Bulletin de la Société française de photographie*, **22** (April 1876), 100–6.
15. Janssen, Jules, "Présentation d'un spécimen de photographies d'un passage artificiel de Vénus obtenu avec le revolver photographique", *CRAS*, **79** (06 July 1874), 6–7.
16. Leclerc, Joseph, "Jules Janssen et le cinématographe", *l'Astronomie*, **69** (1955), 388–91. Tosi, Virgilio, *Les origines du cinéma scientifique 1*, Paris, CNRS broadcast, 1997.
17. Mannoni, Laurent, *Trois siècles de cinéma*, Paris, Éditions de la Réunion des musées nationaux, 1995.
18. Flammarion, Camille, "Le passage de Vénus. Résultats des expéditions françaises", *La Nature*, No. 101, 08 May 1875, 356–8.
19. Canales, Jimena, "Photogenic Venus. The 'cinematographic turn' and its alternatives in the nineteenth-century France", *ISIS*, **93** (2002) 583–613.
20. Letter from George Airy to John Dallmeyer, Cambridge University Library (CUL) RGO Ms 6/277, 327.
21. Letter from Jules Janssen to Warren De La Rue, 11 June 1873, CUL RGO Ms 6/277, 769.
22. Letter from Jules to Henriette Janssen, 06 October 1873, BIF Ms 4133–180.
23. Letter from George Airy to Warren De La Rue, 25 October 1873, CUL RGO Ms 6/277, 774.
24. De La Rue, Warren, "On a piece of apparatus for carrying out M. Janssen's method of time-photographic observations of the transit of Venus", *MNRAS*, **34** (1874), 347–53.
25. Launay, Françoise, "Les acteurs de la mission de Jules Janssen au Japon en 1874", *l'Astronomie*, **121** (2007), 94–101.
26. *L'Illustration*, 25 July 1874; *L'Univers illustré*, 08 August 1874.
27. Letter from Félix Tisserand to Henry Léauté, Anne Guigan-Léauté archives.
28. Letter from Henriette Janssen to the permanent secretary, 26 September 1874, *CRAS*, **79** (09 November 1874), 1064–5.
29. Janssen, Jules, "Lettre de M. Janssen à M. Dumas, président de la Commission du passage de Vénus", *CRAS*, **80** (08 February 1875), 342–5.
30. Letter from Jules Janssen to Jean-Baptiste Dumas, 03 November 1874, AAS, J.-B. Dumas archive, Janssen file.
31. Navy historical archives, Picard file.
32. Letter from Félix Tisserand to Henri Léauté, Anne Guigan-Léauté archives.
33. Janssen, Jules, "Présentation du revolver photographique et épreuves obtenues avec cet instrument", *Bulletin de la Société française de photographie*, 22 (avril 1876), 100–6.
34. Dick, Steven J., Orchiston, Wayne, and Love, Tom, "Simon Newcomb, William Harkness and the nineteenth century American transit of Venus expeditions", *JHA*, **29** (1998) 221–55.
35. Janssen, Jules, "Lettre de M. Janssen à M. Dumas, président de la Commission du passage de Vénus", *CRAS*, **80** (08 February 1875), 342–5.
36. Janssen, Jules, "On the photographic revolver, and on the observations of the transit of Venus made in Japan", in *Report of the 45th Meeting of the BAAS held at Bristol in August 1875*, Murray, London, 1876, p. 26.
37. Débarbat, Suzanne and Launay, Françoise, "The 1874 Transit of Venus observed in Japan by the French, and associated relics", *Journal of Astronomical History and Heritage*, **9** (2006), 167–71.
38. Letter from Jules Janssen to Jean-Jacques Henner, Henner Museum archives, No. 11.
39. Letter from Jules Janssen to Jean-Baptiste Dumas, 11 March 1875, AAS, J.-B. Dumas archives, Janssen file.
40. Janssen, Jules, "On the total eclipse of April 1875, observed at Gangshallô (Siam)", in *Report of the 45th held at Bristol in August 1875*, p. 24.

41. Letter from Jules to Henriette Janssen, 12 September 1876, BIF Ms 4133–194.
42. Janssen, Jules, "Notes sur divers points de physique céleste", *CRAS*, **96** (26 February 1883), 527–9.
43. Janssen, Jules, "Notes sur les passages des corps hypothétiques intra-mercuriels sur le Soleil", *CRAS*, **83**, (02 October 1876), 650–5.
44. Janssen, Jules, "Présentation du revolver photographique et épreuves obtenues avec cet instrument", *Bulletin de la Société française de photographie*, **22** (April 1876), 100–6.
45. Braun, Marta, "The expanded present: photographing movement", in *Beauty of another order. Photography in Science*, Ann Thomas, ed., Yale University Press, New Haven and London, 1997, 150–84.
46. Marey, Jules, "Sur la reproduction, par la photographie, des diverses phases du vol des oiseaux", *CRAS*, **94** (13 March 1882), 683–4.
47. Janssen, Jules, "Remarques sur la communication de M. Marey", *CRAS*, **94** (13 March 1882), 684–5.
48. Janssen, Jules, "Note sur le principe d'un nouveau revolver photographique", *CRAS*, **94** (03 April 1882), 909–11.
49. Marey album, Collège de France archives.
50. Janssen, Jules, "Analyse chronométrique des phénomènes électriques lumineux", in *Report of the 58th meeting of the BAAS held at Bath in September 1888*, 615–6.
51. Millochau, Gaston, "Contribution à l'étude de la décharge intermittente" *CRAS*, **142** (26 March 1906), 781–3.

Chapter 8
The Foundation of the "Paris Observatory for Physical Astronomy" ... Located in the Meudon Estate

"Puisque l'Astronomie physique ne peut plus se confondre désormais avec l'Astronomie mécanique, donnons-lui un établissement séparé".

[Because physical Astronomy can, henceforth, no longer be amalgamated with mechanical [dynamical] *Astronomy, let us give it a separate establishment.]*

Hervé Faye [1]

Janssen's Request

On his return from India in July 1869, and with the ribbon of the Légion d'honneur in his buttonhole, Janssen had undoubtedly acquired the reputation necessary to request further help from Victor Duruy, who had supported him to such an extent in previous years as head of his ministry. He met him very frequently in 1869 and 1870, but it was no longer a question of asking him for finance to mount an expedition. The object was on a completely different scale: nothing less than the creation of a new observatory, dedicated to celestial physics! It was in November 1869 that Janssen sent a first version of his project to the emperor. In particular, he pointed out:

> Over the last ten years, great discoveries have been made about light. These discoveries, when applied to astronomy, have given results of the greatest significance and which have been utterly unexpected. Astronomy now has another branch, and this branch is already so considerable that it merits being provided with special means of observation [2].

The situation appeared to be particularly favorable following the decisions taken several months earlier, about which Henriette informed her husband:

> The Académie has just finished the great debate about the question of the transfer of the Paris Observatory. The Paris Observatory will remain in Paris and remain a first-class observatory. An Observatory, also First-Class, and completely independent of the former, will be created at Versailles or at Fontenay-aux-Roses. This is what is printed in the journal "Les Mondes" today, 9 April *[1869]*. – Will they perhaps offer you some post at this observatory? [3]

F. Launay, *The Astronomer Jules Janssen: A Globetrotter of Celestial Physics*, Astrophysics and Space Science Library 380, DOI 10.1007/978-1-4614-0697-6_8,

And several days later she adds:

> The annual meeting of the Académie has not yet taken place, which leads me to hope that perhaps you may be able to attend. It is undoubtedly all the discussion about the question of the transfer of the *[Paris]* Observatory that has delayed the date. You know from my last letter what the outcome has been. I do not know if I delude myself, but because this *[new]* Observatory will have been authorized, above all, by the Minister's advice following the requests by MM. Faye, Delaunay, etc., and because these gentlemen, as well as the Minister, are extremely well-disposed towards you, and appreciate your work and your gift for invention and observation, but I imagine that they will create a position for you at the *[new]* Observatory. In any case, M. Deville[1] has reserved one (a minor one perhaps) at his Meteorological Observatory [4].

In December 1869, Janssen attended the Minister's Grand Audience [5], with Faye's blessing [6], but a note from the ministry dated 19 November indicates that although the agreement had been reached regarding the right of physical astronomy to exist, the same could not be said for the creation of a new establishment [7]. It is true that Duruy had no longer been Minister since 18 July. In May 1870, however, Janssen was assigned the Pavillon de Breteuil,[2] but France declared war on Prussia and, at the end of August, Janssen returned to rue Labat.

Agreement by the Académie des Sciences

The special role that Janssen played in the following months in relation to the Government of National Defense, presided over by Trochu, and in particular with Jules Simon, and Gambetta, and then with the head of the executive power, Adolphe Thiers, clearly illustrates he had the ear of those at the highest level in the political world, and not simply through his involvement with aerostation. He drew up another document on 15 September 1870 [8]. The markers were being laid down, but the time was not favorable for the outcome of the project. It was to gather information about it that he was to benefit in July 1871, with his fact-finding trip to Britain. He explained then to Jules Simon how much he had been "struck by the enthusiasm and by the sacrifices that our neighbors are making for the progress of science" [9], citing "Monsieur Huggins, our model in celestial physics, *[who]* has just been provided with a magnificent instrument to allow him to follow up his discoveries" and "the Kew Observatory, which has just [...] received an additional allocation of 10,000 pounds sterling".

In June 1873, by which time Janssen had become an academician, it was for a *colleague*, and no longer for someone simply assigned to a mission, that Le Verrier suggested to the minister that he should "confer on this scientist a new post as titular astronomer that would be created at the Paris Observatory" [10]. It should also be

[1] Charles Sainte-Claire Deville.

[2] The Pavillon de Breteuil was an outbuilding of the château de Saint-Cloud (nowadays destroyed), located at Sèvres, near Meudon, South-West Paris. The Bureau international des poids et mesures (BIPM) has been located in it since 1875.

said that for several months Janssen had been one of the six members of the Paris Observatory's Council: his name had been added at the last minute to the decree of 13 February, relating to the second nomination of Le Verrier as head of the establishment [11], after the accidental death of Delaunay, the decree being signed by ... Jules Simon. Unfortunately, Le Verrier received a negative reply from the minister: "I would have been happy to ensure that the Observatory should have the collaboration of M. Janssen, whose very real merit I personally appreciate. But it seems impossible to me to register a new request for an increase that comes to 12,000 F, to the request that has already been submitted to the budgetary commission" [12]. Le Verrier promised Janssen to return to the question "as soon as there is an opportune date" [13], an easy way of postponing it indefinitely. It was definitely very difficult for Janssen to obtain a secure position, because, in May 1872, he had already failed by one vote (in favor of Mascart) in his election as a lecturer at the Collège de France, as replacement for Victor Regnault who had given notice [14]. Janssen did, however, have a steady income (5,000 F per year) following his election to the Bureau des longitudes on 26 May 1873.

It was not until July 1874 that the file concerning the new observatory was reopened, thanks to the intervention of the deputy for Hautes-Alpes, Ernest Cézanne, "the eminent engineer, the principal founder of the French Alpine Club" [15]. Fortified by his support from the Académie des sciences, which had recently elected him to its ranks, Janssen suggested to Cézanne that he should call on the services of the Académie: "What appears to me to be the most suitable *[course]* is for the Académie des sciences to be consulted about the benefit from its formation. I have spoken to Monsieur Dumas, who asks you to be kind enough to see him about the matter" [16], he wrote on 13 May 1874. He composed a final report before his departure for Japan:

> It is here *[in France]* that D'Alembert, Clairaut, Lagrange, Laplace and, more recently, M. Le Verrier have produced magnificent developments in celestial mechanics. Today, all of a sudden, immense horizons are opening up in a new direction, and another astronomy, perhaps even more sublime than its predecessor, is rapidly forming. France cannot remain uncommitted to it [17].

The minister was quickly convinced, and the Académie des sciences set up a committee, consisting of Lœwy, Edmond Becquerel, Bertrand, Dumas, and Faye who acted as rapporteur. While Janssen had his rendezvous with Venus, in November, Faye (Fig. 8.1) presented a report that was not just favorable, but was enthusiastic:

> ... polarization and spectral analysis, especially the latter, have given physical astronomy a completely new impetus; they open immense prospects for stellar astronomy [...]; but what they allow us to glimpse is so beautiful, so grand, and so unexpected that never before has philosophical curiosity had such a stimulus to advance, whatever the price might be. That is why we see great efforts being made all around us. In Italy a Society for spectroscopists has been founded; in England they have created physical chemistry observatories for M. Huggins and for M. Lockyer [...]; in Berlin, they are establishing a solar Observatory for M. Kirchhoff, alongside the astronomical Observatory; in the United States [...] a vast array of work has been organized; here by M. Rutherfurd, there by Dr Young, and elsewhere by M. Langley. And this is quite natural, because the Sun is forced, by a physicist's humble instrument, to shed the veil over its splendor and to reveal, at any time, mysteries that formerly astronomers could only glimpse thanks to rare total eclipses [18].

Fig. 8.1 *Hervé Faye.* Album Félix Potin. © Author's collection

How best could one speak of the glories of spectral analysis and the two scientists who knew so well how to use it in astronomy, the two medalists Janssen and Lockyer? And what a plea he then makes in favor of the "young branch of the old Astronomy"!:

> Here it is no longer Geometry and Mechanics that dominate, it is Physics and Chemistry. Everything in these new Observatories, both instruments and personnel, has been forced to take a special form. Is personnel involved? We find physicists, such as M. Huggins and M. Miller, or M. Lockyer, sometimes combined with chemists such as M. Frankland. Is it a question of instrumentation? No longer do we have meridian circles [...] one might say that it is a chemical laboratory instead [19].

In his lyrical flight, Faye reaches extremes of pathos:

> So it should come as no surprise, nor as a cause for alarm for this new branch of astronomy should be detached from the old trunk. It has taken root in its own soil and is growing rapidly. We older astronomers have difficulty in seeing where we are, given that the ideas, the methods, the objects that are in mind, and even the atmosphere that prevails, differ from our own. [...] Because physical Astronomy can, henceforth, no longer be amalgamated with mechanical *[dynamical]* Astronomy, let us give it a separate establishment: the two sciences will thus develop in parallel, without interfering with one another, *using differing abilities* [20].

Some days later, Flammarion sent a strong endorsement to the journal *L'Événement*, which enabled him, in passing, to settle some old scores with the director of the Paris Observatory:

> The Académie has adopted the conclusions of this report. We may, therefore, hope that soon a new observatory will arise in Paris or its surroundings, provided with new instruments which will allow us to expand interesting research in physical astronomy, which cannot be carried out in the establishment directed by M. Leverrier *[sic]*, and which are quite insufficient when one is reduced to one's own resources. Thank you to the Académie. At last French astronomers understand that the divine science of the heavens should no longer be restricted to showing us inert stones moving in space, nor to tables of logarithms on Earth [21].

The Decree of 1875

The battle had been won, but concrete expression of the outcome had to wait, and, during the summer of 1875, Henriette started to despair: "I am longing for you to return, that you might get them to deliver the decree: as long as I hear nothing said, it seems to me that this observatory is going to escape us; and then I am in a hurry to see you at ease, able to carry out your research work, which provides rest for your mind and ideas" [22], she wrote, on 6 September, to her husband, who was attending the BAAS meeting in Bristol.

And it was on that self-same 6 September 1875 that Mac-Mahon, the President of the French Republic, signed Decree No.4668, which created, at Paris, an observatory for physical astronomy, whose Director answered exclusively to the Minister for Public Instruction, Worship and Fine Arts. The Minister, Henri Wallon, countersigned the decree, which was accompanied by an allocation of 50,000 F to meet the initial expenses for premises and maintenance.

Also named in the decree, the Director of this new observatory was, of course, Jules Janssen (Fig. 8.2). He was 51. He was a graduate of neither the École normale supérieure, nor of the Polytechnique and, until then, had only held temporary assignments from the state. He finally saw the outcome of 18 years of research in the service of science, and the establishment created for him blithely avoided the reform of French observatories that had just been put into effect, and where the decree of 1873 [23], which laid down the organization of the observatories at Paris, Marseille, Toulouse, and Algiers (the only public, French observatories at that time), precisely did not apply to the new observatory!

But where would the establishment be located?

Fig. 8.2 *Portrait of Jules Janssen from a photograph by Paul Berthier.* L'Univers illustré, *08 August 1874.* © Author's collection

The Choice of Meudon

In a report written in May 1874, Janssen had given precise details of the ideal situation:

> The Observatory should be located at a site that is relatively open, primarily on the southern side and to the east, far from vibration and from the illumination of the atmosphere produced by gas. It is also desirable that the Observatory should not be located in the middle of a forest, nor in too extensive a wood but, as far as possible, on a slight rise surrounded by meadows, turf, and generally low vegetation. Easy connections with Paris are absolutely indispensable. It is therefore desirable that the site adopted should be located at the *[shortest]* distance that is strictly necessary to avoid the capital's ground vibrations and the atmospheric illumination [24].

In 1896, in the first volume of the *Annales de l'Observatoire d'Astronomie Physique de Paris* where he indicated that "the observatory was provisionally set up on the boulevard d'Ornano, at the same site where the preparations for the expedition to Japan were made", Janssen explained that he was given the choice between la Malmaison and Meudon. He preferred Meudon, because "this beautiful estate, where the chateau had, it is true, burned down after the war, had been assessed, in the liquidation accounts, at a derisory sum, and if it were to have been sold, would have been divided up and torn down. There was thus question of national interest in retaining such a beautiful estate for the state ..." [25].

So it was no longer a question of Versailles or Fontenay-aux-Roses, and Janssen passed over in silence the suggestion for the location at Vincennes, mentioned in the journal *La Nature* in 1875 [26], as well as the fact that at the War Ministry, which was trying to find a site for him in 1876, Mont Valérien had been mentioned: "I saw General Rivière [*sic*], who received me very kindly, and would even have been prepared to have made space at Mont Valérien. I am leaving for Versailles to see an officer whom he suggested, to see what could be done at Meudon [27] (Fig. 8.3).

Routine Observations of the Sun and the Rumford Medal

It was on 13 October 1876 that Janssen settled into the Communs (outbuildings) on the Meudon estate. In the park (which was still occupied by the army), he then set up his instruments, in particular the photographic solar refractors [28] and the huts that had been returned from Japan: "As the army abandoned Meudon to occupy the new barracks that had been constructed for it, we were able to spread ourselves out and even make use of some of the huts that it had left" [29] (Fig. 8.4). The refractor that gave images 30 cm in diameter, in particular, was mounted inside a dome, 7.50 m in diameter. With the photographer Arents, now taken on at Meudon, Janssen immediately began to take routine photographic records of the Sun, which he considered, quite rightly, to be essential:

> The solar archives have as much significance as human historical archives, because whereas the latter concern the politics of nations, the former lay the foundation for Man's knowledge of the Universe, with all the fruits of intellectual power and pleasure that will flow from it [30].

Fig. 8.3 *The Château de Meudon, blocked up after the fire of 1871.* © Paris Observatory Library

In the December *[1876]*, Janssen was in London to receive a prestigious award, the Rumford Medal[3]:

> So, it is over: I have just received the medal or rather the two medals. A great meeting at Burlington House, where the Royal Society has been housed by the government. I have just seen Warren de la Rue, Hooker, Huxley, Stokes, Lockyer, and Huggins, where I have spent the day. Everyone speaks to me about you, because my travels always call attention to my courageous companion. Claude Bernard has received the other medal[4] that the Society issues, and it has been much pointed out that it is two Frenchmen *[who]* are the only ones honored this year. I told you that there were two medals. The gold one contains some 1,500 F-worth of the metal, and so it is extremely heavy in my pocket. The silver medal is of the same size, but much lighter because of the m*[uch]* lower density of that metal. – The award also includes a sum of £60, I am told. This evening there is a grand dinner in my honor, and already more than one hundred people have registered. A formidable speech is expected from your servant, and I have nothing prepared. You will, therefore, allow me, Madame, to think about it a little [31].

---

[3] The Rumford Medal (in fact two medals), awarded biennially for "an outstandingly important recent discovery in the field of thermal or optical properties of matter made by a scientist working in Europe".

[4] The Copley Medal, a silver-gilt medal awarded annually for "outstanding achievements in any branch of science", the oldest and most prestigious of the Royal Society's medals.

Fig. 8.4 *Janssen's first installations at Meudon.* L'Illustration, *17 October 1885.* © Paris Observatory Library

Lockyer, who had informed Janssen of his honor by telegram, had received the Rumford Medal 2 years earlier, but Huggins had yet to receive it, and we can see that Janssen, the new observatory director, finds that the honors that he receives to be perfectly normal. The gold medal weighed down his pockets for 20 years, until he donated it to the fund destined to help travelers[5]:

> As for that medal [...] I would say to you that, for me, it is the object of a sort of remorse. I am cross with myself for not having thought of this practical use *[for it]*. [...] I was held back by a scruple. It seemed to me that I should leave, within my family, this token of respect for one of its members. But Science, Gentlemen, has sovereign rights, and to serve it, we should silence all other considerations of interest or sentiment [32].

As far as the Copley Medal is concerned, Janssen explained that during the ceremony it was given to the French ambassador, because Claude Bernard was

[5] The members of the society called "Réunion des voyageurs français".

unable to travel to receive it. Janssen himself possibly carried it back to Paris, because the two men knew one another well: they had met for a long time at the Société philomathique before becoming colleagues at the Académie des sciences.

The Law of 1879

On 15 April 1879, Jules Grévy, the President of the Republic, promulgated the law which assigned "the château de Meudon, the Communs of the said château, the orangery, etc., with part of the park […] and the avenue de Bellevue, to the establishment of the Paris Observatory of Physical Astronomy. […] The costs for the repair of the buildings and for fencing the *[parts of the]* park assigned to the observatory, as well as the cost of acquiring the instruments and furniture are set at the sum of one million and thirty five thousand francs, to be charged in three parts to each of the financial years 1879, 1880, and 1881" [33]. Jules Ferry, Minister for Public Education and Fine Arts (Religion being now under the Interior Ministry), countersigned the bill, together with his colleagues for Public Works and Finance. It may be noted that in the Senate, the rapporteur of the committee charged, in 1878, with examining the law was the Alsatian chemist and industrialist Auguste Scheurer-Kestner, a great friend, like his friends Gambetta and Jules Ferry, of Janssen's great Alsatian friend, the painter Jean-Jacques Henner …

Janssen immediately sent the text of the law to Lockyer, and confirmed support for him to have an observatory more suited to his work than the one at South Kensington:

> I repeat here what I have said many times, that it is highly desirable, not only just as a form of debt owing to you, but also for the reputation of distinguished English science that an observatory should be constructed for you, and which enables you to engage, without worries, and with suitable means, in pursuing your wonderful work [34].

Lockyer did, in fact, rely considerably on the help of his friend Janssen:

> Whether we have a permanent Solar Observatory some day depends upon the answer you and others give to our letter. If you would write and say that you approve our plan & will join with us & point out the importance of a solar observatory we may have one someday. I wish you would come over and stay with me. I am all alone. you would have a warm reception here [35].

Janssen's reply was not long in coming: "If it only rested with me, you would have a solar observatory immediately – and a good one."

At Meudon, it was now possible to begin the work on transforming the charred remains of the "château neuf" that the fire of 31 January 1871 (3 days after the armistice) had devastated over 3 days. The architect Constant Moyaux had carried out a survey in 1878, and the two men had the same lofty ideals in mind. This may be seen from the letter that Janssen sent to Moyaux several years later:

> You have splendidly said that it is love of France just as much as that of science or art that should possess us. To me, that is a sentiment of which I approve more and more since 1870. In their own sphere, no one can set themselves a goal that is more necessary, more urgent, more noble! [36] (Fig. 8.5)

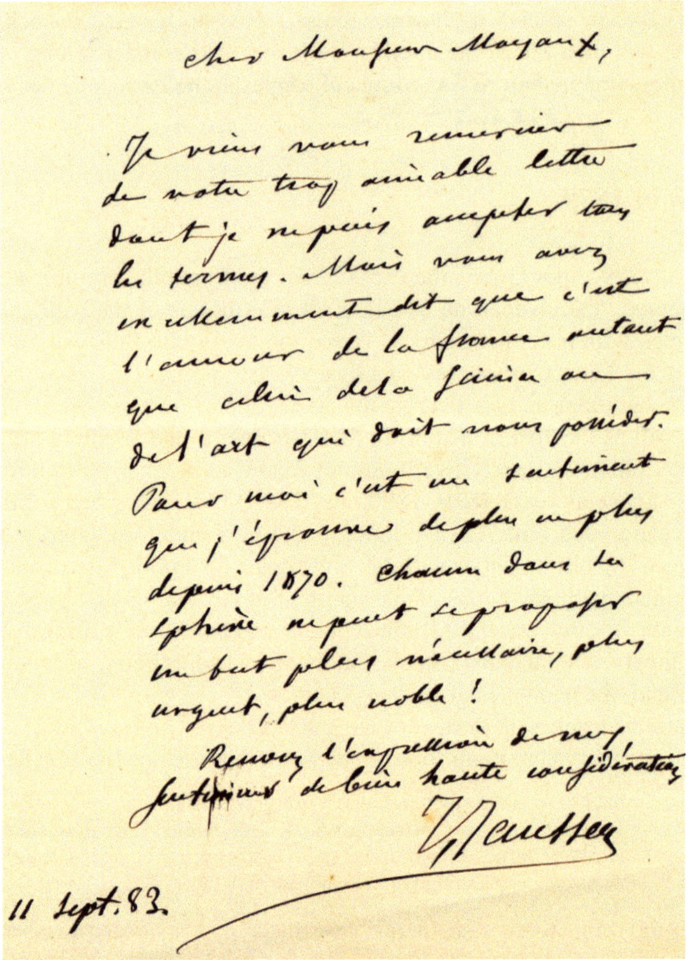

Cher Monsieur Moyaux,

Je viens vous remercier de votre trop aimable lettre dont je ne puis accepter tous les termes. Mais vous avez excellemment dit que c'est l'amour de la France autant que celui de la science ou de l'art qui doit nous posséder. Pour moi c'est un sentiment que j'éprouve de plus en plus depuis 1870. Chacun dans sa sphère ne peut se proposer un but plus nécessaire, plus urgent, plus noble !

Recevez l'expression de mes sentiments de bien haute considération

J Janssen

11 Sept. 83.

Fig. 8.5 *Letter from Jules Janssen to the architect Constant Moyaux.* © Author's collection

The Great 83-cm Refractor

The work was all the more significant and complex because Janssen wanted to house the largest refractor in Europe [37] in the central part of the château, beneath a large dome, with a visual objective, 83 cm in diameter, corrected in the yellow–green region, and a photographic objective, 62 cm in diameter, corrected for the blue–violet. Indeed, at the time, only the Lick refractor, commissioned in 1888, had a larger diameter (91 cm). The Meudon refractor remains today the third in the world, after the construction of the Yerkes instrument in 1897, with a diameter of 1.02 m. Janssen, evoking Galileo Galilei, liked to say "It was the invention of the refractor that gave physical astronomy its basis" [38].

Fig. 8.6 *The great Meudon refractor.* Annales de l'Observatoire d'Astronomie Physique de Paris. © Author's collection

The construction of the equatorial mounting was entrusted to the Gautier company from the boulevard Arago. This mounting carries the double refractor, with its focal length of 16 m: "The two instruments are carried within a single tube [...] consisting of seven steel sleeves [...]. The whole forms a very rigid, rectangular tube" [39] (Fig. 8.6).

The observatory council examined the plans for the transactions on 15 November 1886 [40]. The cost of the objectives amounted to 130,000 F and 50,000 F for the visual and photographic objectives, respectively. Like the objectives for the giant American refractors, the glass was supplied by the Paris factory of Édouard Mantois, who went into partnership with Charles Feil, the grandson of Henri Guinand, in 1885. And it was the two great French opticians of the period, Paul and Prosper Henry, who were given the task of surfacing the lenses, in accordance with an agreement that Janssen entered into with them on 27 December 1885 [41].

The contract relating to the construction of the iron dome, with an interior diameter of 18.50 m, whose weight is about 100 tonnes, and which is moved electrically – a fact on which Janssen laid great emphasis in his descriptions – was signed in November 1886 with the firm Établissements Cail. The acceptance took place nearly 9 years later, on 10 June 1895, when the refractor had already been finished for some 18 months. In fact, it was precisely on 15 December 1893 that Janssen made his first observation from the great dome: the Pole Star. The following 12 January *[1894]*, it was the turn of the Moon and the planet Jupiter [42]. Indeed, the refractor, which allowed the use of very high magnifications, was particularly suitable for the study of planets and, in December 1896, Janssen stopped his series of observations of Jupiter to allow his new collaborator, Henri Perrotin (the future Director of the Nice Observatory), to observe Mars through the eyepiece during the favorable opposition [43]. In 1897, Henri Deslandres, who began his astronomical career at Paris, was transferred to Meudon. In 1898 he used the photographic refractor to study the Orion Nebula, the planetary nebula in Andromeda, and the one in Draco, as well as a number of variable stars in various clusters. This was occasion for Janssen to remark that "instruments with very long focal lengths, and thus provided with considerable powers of separation *[are]* valuable for the study of the structure of very brilliant and small-diameter stars", and took delight in "the quality of the objectives constructed by MM. Henry" [44]. Deslandres later developed the technique of photographic spectroscopy with the great refractor. In 1904, Janssen congratulated himself for having prompted and encouraged photographic studies of the spectrum of the planet Jupiter, with the help of the astronomer Gaston Millochau [45], who came from Paris with Deslandres.

Despite the excellent work done by the great refractor in Janssen's time, and during the last century, and which it may still carry out (for the study of planetary surfaces and double stars, in particular, but also for in public observation sessions), it is obvious that, in 1896, the reign of large refractors was coming to an end. They would never be made any larger than the 1.25 m diameter objectives constructed for the giant refractor made for the Universal Exhibition of 1900 and which, given its length of 60 m, took the form of a horizontal tube, with starlight being directed into it from a 2 m-diameter mirror on a giant Foucault siderostat [46]. We now know that, regrettably, the prowess exhibited by the constructors, both the opticians and the mechanical engineers, was overshadowed by the financial bankruptcy of the project, and that the silent hostility of official scientific circles was such that this dream instrument never found a home in any observatory, once the Exhibition had finished. Nevertheless, it does seem from the transport costs recorded in his notebooks that Janssen considered giving it a home at Meudon [47]. Janssen had, indeed, been a contributor to the Palais de l'Optique venture, for which he had provided photographs of the Sun.

The Large 1-m Reflector

Although Meudon's large refractor, an instrument that almost has a mythology of its own, nowadays appears as a "fin de siècle" instrument, the extremely fast reflector, 1 m in diameter [48] and with a very short focal length (just 3 m), which was finished at the same time, was, on the other hand, a new-generation instrument. It did, in effect, initiate the promising era of large reflectors, which is still, nowadays, at the beginning of the twenty-first century, far from over. The refractor was undoubtedly

Fig. 8.7 *The 1-metre reflector at Meudon.* Annales de l'Observatoire d'Astronomie Physique de Paris. © Author's collection

just one element in the race for colossal size, whereas the reflector was a completely new tool. Producing a silvered glass, parabolic mirror was also one of the achievements of the brothers Henry. It was not a trivial matter, and the mechanical construction was also bold. Indeed, the mounting was designed by Gautier to be transportable:

> The reflector's mounting is built on the English system *[an English mount]*, but where both ends of the polar axis are carried by a strong circular arc, which may be moved [...] relative to the base on which it rests. The arrangement allows the inclination of the polar axis to be altered and, as a result, to use the instrument at sites over a wide range of latitudes. It therefore forms a powerful transportable instrument [49] (Fig. 8.7).

When one knows that the dome that shelters the reflector is a "small" dome with a diameter of 7.5 m, it is obvious that Janssen never had the slightest doubts about anything!

Janssen, who recalled "the success obtained in 1871 with a reflector with a very short focal length" [50] which enabled him to recognize the gaseous nature of the solar corona, did not fail to stress how "valuable this instrument is, on account of its enormous light-gathering power" [51], especially for observing nebulae. It was thus that "M. Rabourdin, who comes from the Observatory at Algiers, has photographed nebulae with exposure times that are four to ten times less than previous studies" [52].

Fig. 8.8 *Janssen in his spectroscopy laboratory.* Photograph. © Audouin Dollfus collection

The Laboratory for the Analysis of Gases

So it was 20 years after he settled at Meudon that Janssen commissioned his observatory's two large, prestigious instruments. The same did not apply to the laboratory for the analysis of gases, which was far less difficult to install in the château's Communs. The latter did, in fact, allow a laboratory 100 m long to be created rapidly, and which could be extended to 140 m. The outbuildings having been converted into stables in the nineteenth century, it was not even necessary to remove the loose-boxes, whose oak partitions formed excellent supports for the 60-m long absorption tubes that Janssen required for his experiments. The tubes, 5–6 cm in diameter, assembled in 6-m lengths, were of copper, and tested to no less than 200 atmospheres! For even higher pressures, steel tubes were used. As a light source, Janssen used either "Drummond light[6]", obtained with pure, compressed oxygen, town gas, and cylinders of calcium oxide *[quicklime]* or magnesia" [53]; electric light obtained from accumulators or from a dynamo; or sunlight. Naturally, he had a large number of spectroscopic instruments, including a very fine spectrometer by Brunner, as well as photographic cameras (Figs. 8.8 and 8.9). This was how Janssen studied hydrogen, nitrogen dioxide, nitrous oxide, and carbon dioxide [54].

[6] More commonly known as "limelight" or "calcium light".

Fig. 8.9 *Janssen's laboratory at Meudon for the analysis of gases.* Annales de l'Observatoire d'Astronomie Physique de Paris. © Author's collection

One tube was erected outside for use when the experiments involved danger, either because of the high pressures employed or because of the nature of the gas being studied. In addition, a special installation consisting of tubes of steel was used when the gas had to be taken to a high temperature (up to 900°C), but that still did not prevent certain incidents from taking place. Albert Nodon, who was associated with these studies in the rôle of observatory physicist, recounts:

> I also felt that I ought to describe an experiment that Janssen had set up with a great deal of care to study the spectrum of oxygen under pressure and at a high temperature; an experiment that nearly caused a catastrophe, and which, besides, has never been repeated later. [...] I had helped him in setting up the apparatus, which consisted of a solid tube of steel, standing vertically, and closed at both ends by thick glass. Through the whole length of this tube there ran a platinum spiral [...]. This tube contained pure oxygen, at a pressure of several hundred atmospheres, which was expected to be heated to incandescence by the glowing spiral. [...] We were in place at the spectroscope: the current was sent through the spiral, which rapidly brought to a blinding glow. At the moment when we started to make our observations, a violent explosion shook us, and we found ourselves surrounded by a blinding flash. [...] Rather taken aback by this abrupt end to the experiment, and happy to have come out of it so well, because we had not been hurt, we immediately found ourselves surrounded by the whole personnel of the Observatory, attracted by the tremendous noise of the explosion. Janssen, who preserved a supreme control over himself, whatever the situation, solemnly descended the stairs, without any apparent emotion, turned to me and said "We will have to start again, let's go and have lunch" [55].

Fig. 8.10 *Photograph of Paris under the snow, taken from a wing at Meudon by Janssen on 18 December 1878.* © Paris Observatory Library

The Director's Accommodation

When he settled at Meudon in 1876, Janssen, who had achieved his life's main goal, because thenceforward he had a "position", divided his time between his laboratory research, his studies of the Sun, his instrumental projects, his travels, his contribution to meetings…

He lived in the Communs, from which he enjoyed a fine view over Paris. On 16 December 1878, he took a photograph of the capital under the snow from his kitchen, which was in the eastern wing of the second courtyard (Fig. 8.10), and Henri Dehérain reported that his living quarters contained souvenirs of his travels: fabrics, vases, lacquer plates, and small ebony elephants from India [56]. Some of these curiosities are visible in the engraving that appeared in 1895 in *La Nature* [57], which shows Janssen in his drawing room at Meudon (Fig 8.11). Apart from the souvenirs there are some works by artists: a bust of Janssen by Tournois, another of Mme Janssen, and portraits of Janssen by Henner and Paul Dubois. We may add that we can also see a large photograph of Mont Blanc by Tairraz, the photographer from Chamonix and, on the table, the photograph of a total eclipse. In 1905, journalists from the *Petit Journal* noticed, in addition, "engravings that show the scholar's taste for archaeology", a medallion of Gounod on the mantelpiece, and Japanese swords with delicately ornamented scabbards, and razor-sharp blades [which] seem to invite national "harakiri" [58]!

Fig. 8.11 *Janssen in his drawing room at Meudon, after a photograph by M. Dornac.* La Nature, *19 October 1895.* © Author's collection

"Janssen's official office has but a single chair for his personal use" [59], explained Tikhov, who arrived at Meudon in 1897. "Familiar visitors, in particular Observatory collaborators, spoke to him standing up. It was only for eminent visitors, and in particular for foreigners, that he had chairs brought in." He went on to add "The scientist had an imposing presence. Long grey hair framed his large head and its distinguished features. Moustache and beard were white. His overall demeanor was a rarely encountered solemnity."

Janssen did not, however, live permanently at Meudon. Indeed, because he needed to be in Paris very frequently to attend the meetings of all the societies (at least 15) to which he belonged, and which he was not in the habit of missing if he was not on a trip to the provinces or abroad, he had a pied-à-terre in the capital: 63 rue de Vaugirard (at the junction with the rue de Rennes), from at least 1886 onwards. He explicitly mentions this in his notebooks, and it was there that, 1 day in April 1889, he received a writ [60], from the optician Prazmowski's successors, demanding payment or restitution of an objective, valued at 10,000 F (his annual salary as director), which had been entrusted to him to try. As, added to that, there was a debt of 1,005 F for various items supplied, one cannot help thinking of the two doors to Balzac's house, which allowed the latter to escape, occasionally, from his debtors!

This second address also explains why there were two religious services before Janssen was buried at Père-Lachaise Cemetery: the first at the Église Saint-Martin at Meudon, and the second at Saint-Germain des Prés, his Parisian parish (Fig. 8.12).

M

Madame Jules Janssen,
Mademoiselle Antoinette-Marie Janssen,
Le Lt Colonel Sénèque de Blémont, Officier de la Légion d'Honneur, et Madame Sénèque de Blémont, Monsieur Ernest Sénèque de Blémont, Mademoiselle Eulalie Sénèque de Blémont, Monsieur et Madame Charles Schreiber et leur fille,
Monsieur Eugène Janssen, ancien Banquier, Monsieur Amédée Loiseau, Inspecteur Principal Honoraire de la Cie d'Orléans, Chevalier de la Légion d'Honneur, et Madame Amédée Loiseau, Monsieur Henry Le Roux, Directeur Honoraire des Affaires Départementales de Préfectures, Officier de la Légion d'Honneur, Madame Hippolyte Dufort, Madame Clémence Forestier, en religion Sœur Béatrix du Bon Pasteur, Supérieure du Couvent du Refuge, Madame Pouillart, ses enfants et petits-enfants,
Madame Eugène Labiche et ses enfants, Monsieur Raoul Bompard Ancien Député, Conseiller de la Préfecture de la Seine, Madame Raoul Bompard et leurs enfants, Monsieur et Madame Maurice Dufort et leurs filles, Monsieur et Madame Henri Oriol, Monsieur Albert Dufort, Ingénieur Civil, Monsieur Georges Dufort, Attaché au Consulat de France à Gênes, Mademoiselle Blanche Grandvallet, Mesdemoiselles Fanny et Berthe Lionnet,
Les Elèves et le Personnel de l'Observatoire de Meudon,
Ont la douleur de vous faire part de la perte cruelle qu'ils viennent d'éprouver en la personne de

Monsieur Pierre, Jules, César Janssen,

Doyen de l'Académie des Sciences,
Membre du Bureau des Longitudes,
Directeur et Fondateur de l'Observatoire d'Astronomie Physique de Meudon
et de l'Observatoire du Mont-Blanc,
Président d'Honneur de la Société de Géographie, de la Société Astronomique de France,
de la Société de Physique, de la Commission Permanente Internationale d'Aéronautique,
de l'Union des Sociétés Photographiques de France, du Club Alpin Français et de la Société d'Hygiène,
Membre de la Société Royale de Londres, de l'Académie dei Lincei, de l'Académie des Sciences de Palerme,
Associé étranger de l'Académie des Sciences des États-Unis,
Membre Correspondant de l'Académie des Sciences de Saint Pétersbourg,
de la Société des Spectroscopistes d'Italie,
Associé de la Société Royale d'Astronomie de Londres, de l'Académie Royale de Belgique
et de l'Académie des Sciences de Bonn,
Membre Honoraire de la Société de Physique de Londres, de la Société Helvétique de Lauzanne,
et de l'Académie de Dublin,
Docteur des Universités d'Edimbourg et de Dublin,
Commandeur de la Légion d'Honneur,
Dignitaire de l'Ordre de la Rose du Brésil,
Commandeur de l'Ordre de Saint Jacques de l'Épée du Portugal,
Grand Croix des Ordres de Saint-Stanislas (1re Classe), de Russie,
et de Saint Sava (1re Classe), de Serbie,
Grand Officier de la Couronne d'Italie,

leur époux, père, oncle, grand-oncle, cousin et petit-cousin, décédé à l'Observatoire de Meudon, le 23 Décembre 1907, dans sa 84e année, muni des Sacrements de l'Eglise.

Cœli enarrant Gloriam Dei.

Deux Services Religieux ont été célébrés dans les Eglises Saint-Martin de Meudon et Saint-Germain-des-Prés, à Paris.

L'Inhumation a eu lieu au Cimetière du Père Lachaise.

Observatoire de Meudon.

Administration Spéciale des Funérailles et des Transports Funèbres, Maison Henri de Borniol, 28, Rue de la République à Meudon-Bellevue (Seine-et-Oise), et 88, Rue de Paris à Clamart (Seine).

Fig. 8.12 *Announcement of Jules Janssen's death.* © Paris Observatory Library

References

1. Faye, Hervé, "Rapport de la Commission nommée le 17 Août pour préparer une réponse à la lettre adressée par M. le ministre de l'Instruction publique, au sujet de l'opportunité de la création d'un observatoire d'Astronomie physique aux environs de Paris", *CRAS*, **79** (02 November 1874), 1018–24.
2. Memorandum from Jules Janssen addressed to the Emperor, November 1869, AN F[17] 3745.
3. Letter from Henriette to Jules Janssen, 09 April 1869, BIF Ms 4134–100.
4. Letter from Henriette to Jules Janssen, 16 April 1869, BIF Ms 4134–101.

5. Jules Janssen's notebook, BIF Ms 4127–9.
6. Letter from Hervé Faye to Jules Janssen, December 1869, BIF Ms 4135–145.
7. Note from the ministry, dated 19 November 1869, AN F^{17} 3745.
8. Jules Janssen's notebook, BIF Ms 4127–10.
9. Letter from Jules Janssen to the Minister Jules Simon, 22 July 1871, AN F^{17} 2977.
10. Letter from Urbain Le Verrier to Jules Janssen, 14 July 1873, BIF Ms 4136–351.
11. Letter from Jules Simon to Urbain Le Verrier, 13 February 1873, Paris Observatory Library, Ms 1047, D13.
12. Letter from Urbain Le Verrier to Jules Janssen, 14 July 1873, BIF Ms 4136–351.
13. *Ibid.*
14. Draft letter from Jean-Baptiste Dumas to the Minister for Public Instruction, 08 May 1872, AAS, Janssen file.
15. Janssen, Jules, "Création de l'observatoire" in *Annales de l'Observatoire d'Astronomie Physique de Paris*, Vol.1, Gauthier-Villars, Paris, 1896, 49–55, p.51.
16. Letter from Jules Janssen to Ernest Cézanne, 13 May 1874, AN F^{17} 3745.
17. Janssen, Jules, report to the Minister, May 1874, AN F^{17} 3745.
18. Faye, Hervé, "Rapport de la Commission nommée le 17 août pour préparer une réponse à la lettre adressée par M. le ministre de l'Instruction publique, au sujet de l'opportunité de la création d'un observatoire d'Astronomie physique aux environs de Paris", *CRAS*, **79** (02 November 1874), 1010–24.
19. *Ibid.*
20. *Ibid.*
21. Flammarion, Camille, in *L'Événement*, 18 November 1874, AN F^{17} 3745.
22. Letter from Henriette to Jules Janssen, 06 September 1875, BIF Ms, 4134–160.
23. Decree No. 1944 of 13 February 1873.
24. Janssen, Jules, Report to the Minister, May 1874, AN F^{17} 3745.
25. Janssen, Jules, "Création de l'observatoire" in *Annales de l'Observatoire d'Astronomie Physique de Paris*, Vol. I, Gauthier-Villars, Paris, 1896, 49–55, p.52.
26. "Création d'un nouvel observatoire d'astronomie physique", *La Nature*, No. 121, 25 September 1875, 269-70.
27. Letter from Jules to Henriette Janssen, 02 September 1876, BIF Ms 4133-190.
28. Le Cocguen, Régis & Launay, Françoise, "La lunette photographique solaire de Janssen', *l'Astronomie*, **119** (2005), 71–5.
29. Janssen, Jules, "Création de l'observatoire" in *Annales de l'Observatoire d'Astronomie Physique de Paris*, Vol. I, Gauthier-Villars, Paris, 1896, 49–55, p.53.
30. Janssen, Jules, "La photographie céleste" in *AFAS, Compte rendu de la 16e session, congrès de Toulouse (1887)*, 1888, 395–409.
31. Letter from Jules to Henriette Janssen, 05 December 1876, BIF Ms 4133–196.
32. Janssen, Jules, "Allocution prononcée au vingt-deuxième dîner de la Réunion des voyageurs français, le 23 mars 1896", in *Œuvres scientifiques*, Vol. II, 410–1.
33. *Journal officiel de la République française* for Wednesday, 16 April 1879.
34. Letter from Jules Janssen to Norman Lockyer, 19 October 1879, EUL.
35. Letter from Norman Lockyer to Jules Janssen, 01 December 1881, BIF Ms 4136-360.
36. Letter from Jules Janssen to Constant Moyaux, 11 September 1883, Françoise Launay collection.
37. Dollfus, Audouin, *La grande lunette de Meudon*, CNRS éditions, Paris, 2006.
38. Janssen, Jules, "L'astronomie physique" in *Annales de l'Observatoire d'Astronomie Physique de Paris*, Vol. I, Gauthier-Villars, Paris, 1896, 1–46.
39. Janssen, Jules, "Description de l'Observatoire", in *Annales de l'Observatoire d'Astronomie Physique de Paris*, Vol. I, Gauthier-Villars, Paris, 1896, 67–87.
40. AN F^{17} 3745.
41. Letter from Jules Janssen to the Minister, 27 December 1885, AN F^{17} 3745.
42. Jules Janssen's notebook, BIF Ms 4131-3.
43. Janssen, Jules, "Remarque sur une note de M. Perrotin intitulée 'Sur la planète Mars'", *CRAS*, **124** (16 February 1897), 346–7.

44. Janssen, Jules, "Remarques sur une communication de M. H. Deslandres relative aux photographies stellaires obtenues avec la grande lunette de l'Observatoire de Meudon", *CRAS*, **128** (05 June 1899), 1378–80.
45. Janssen, Jules, "Remarques sur la note de M. Millochau relative à l'étude photographique du spectre de la planète Jupiter", *CRAS*, **138** (13 June 1904), 1478–9.
46. Débarbat, Suzanne & Launay, Françoise, "The objectives of the Great Paris Exhibition Telescope of 1900", *Bulletin of the Scientific Instrument Society*, No. 74 (2002), 22–3.
47. Jules Janssen's notebook, BIF Ms 4132-4.
48. Dollfus, Audouin, "Le grand télescope de Janssen", *l'Astronomie*, **114** (2000), 236–50 and 360–71.
49. Janssen, Jules, "Description de l'Observatoire", in *Annales de l'Observatoire d'Astronomie Physique de Paris*, Vol. I, Gauthier-Villars, Paris, 1896, 67–87.
50. Janssen, Jules, "Remarques sur une note de M. Louis Rabourdin intitulée: 'Sur quelques photographies de nébuleuses obtenues à l'Observatoire de Meudon' et sur la méthode propre à donner des nébuleuses des images comparables", *CRAS*, **126** (31 January 1898), 383–5.
51. Janssen, Jules, "Remarques sur la note de M. Millochau relative à l'étude photographique du spectre de la planète Jupiter", *CRAS*, **138** (13 June 1904), 1478–9.
52. Janssen, Jules, "Remarques sur une note de M. Louis Rabourdin intitulée: 'Sur quelques photographies de nébuleuses obtenues à l'Observatoire de Meudon' et sur la méthode propre à donner des nébuleuses des images comparables", *CRAS*, **126** (31 January 1898), 383–5.
53. Janssen, Jules, "Description de l'Observatoire", in *Annales de l'Observatoire d'Astronomie Physique de Paris*, Vol. I, Gauthier-Villars, Paris, 1896, 67–87.
54. Nodon, Albert, "Portraits de savant, Jules Janssen (1824–1907)" in *La Science moderne*, July 1926, 387–92.
55. *Ibid.*
56. Dehérain, Henri, "Préface", in *Œuvres scientifiques*, Vol. I, v–vii.
57. Tissandier, Gaston, "Photographies de nos contemporains chez eux. M. Janssen", *La Nature*, No. 1168, 19 October 1895, 329–31.
58. *Le Petit Journal*, 05 March 1905, p. 1.
59. Tikhov, Gabriel, *L'énigme des planètes*, Soviet Union, foreign-language editions, n.d., p. 50.
60. Paris Observatory Library.

Chapter 9
Janssen, the Photographic Technician

> *"Aussi n'hésité-je pas à dire que la plaque photographique sera bientôt la véritable rétine du savant".*
>
> *[So, I do not hesitate to say that the photographic plate will soon be the actual retina of the scientist.] (Fig. 9.1)*
>
> Jules Janssen [1]

Venus Was the Turning Point

As we have seen, it was as if Janssen had invented cinematography for Venus, before even taking photographs of his beloved Sun! Was it not only 2 weeks before he presented his first trials of his revolver to the Académie des sciences, that he declared "For me, it was the observation of the Transit of Venus that had attracted my attention more particularly to this field *[photography]*, which is so fruitful and so neglected by us" [2]. This was in June 1874, and Janssen showed his colleagues the first photographs of the Sun, recorded thanks to the special equipment that he had had made for Japan.

We must not, of course, ignore the significance of that phrase "more particularly," because Janssen was not the man to allow any innovation to pass by through indifference. In the notebook into which he copied and annotated a number of communications presented to the Académie des sciences between 1842 and 1868 [3], there were mainly texts relating to the observation of eclipses, where photography already played a large part. Among the most significant, there is the one in which Faye gave an extract from a letter from Secchi concerning the photographic experiments made during the solar eclipse of 8 July 1851:

> When this letter was read, M. Faye recalled a scheme for observation that he had proposed several years ago, and which he hopes to be able to achieve, thanks above all, to recent progress in photography. This plan consisted in observing with a daguerreotype the image of the Sun and sunspots, not through a single imprint but with two imprints made on the same plate [...] 2 min. apart [4].

F. Launay, *The Astronomer Jules Janssen: A Globetrotter of Celestial Physics*, Astrophysics and Space Science Library 380, DOI 10.1007/978-1-4614-0697-6_9,

Fig. 9.1 *Jules Janssen. Photographic plate.* © Paris Observatory Library

Again in 1858, when Faye showed the photographs of the eclipse of 15 March, taken by Porro and Quinet with a refractor with a focal length of 15 m on collodion plates, where the Sun had a diameter of 15 cm, and "where the clarity of the outline and the sharpness of the cusps were outstanding" [5], he emphasized that "already M. Bond, in the United States, and M. Delarue *[sic]* in London, have obtained photographic images of the Moon, stars and planets, using an equatorial and a clockwork mechanism" [6], and states unambiguously: "I would dare to maintain that one could, from today, substitute photography for almost all meridian observations of the planets and fundamental stars." If Janssen did not immediately follow this advice, it was that he did not have the means, but it is obvious that he was very deeply interested in the evolution of photographic techniques, which he had carried out himself from the time of his thesis.

Solar Photography in 1874

As he explained at the time of the lecture on celestial photography given at the meeting of the Association française pour l'avancement des sciences (AFAS), on 23 September 1887 [7], it was Fizeau and Foucault who first succeeded in obtaining a complete photograph of the surface of the Sun. This daguerreotype, taken 2 April 1845, shows sunspots as well as limb-darkening. It was another 10 years before a larger and clearer image, obtained by Reade, showed the "mottled" appearance of the surface. At Faye's prompting, Porro subsequently succeeded in photographing "the most delicate mottling that furrows the limbs of the Sun," and it was from 1858 onwards that, at Sir John Herschel's instigation, De La Rue, at Kew, started a systematic monitoring of sunspots through photography. But these images were not designed to study the fine structure of the photosphere, "the intimate constitution of the solar surface." Despite the feats

Fig. 9.2 *The solar photographic refractors at Meudon.* © Paris Observatory Library

of the American, Rutherfurd, and the German, Vogel, images were too small, so it was far from being possible to capture the granulation,[1] which was already difficult to observe visually, and which intrigued so many astronomers.

In 1861 – a date when he was working with Follin – Janssen indicated that "if one were to examine [...] the *[solar]* spectrum from the photographic point of view, that is to say from the point of view of its action on sensitive chemical substances, one would recognize that it is in its lower region, that is in the blue rays, the violet, and even lower in the invisible portion (which is the region where the rays are called ultra-violet), that the photographic effects are primarily found" [8]. It is obvious that the conditional statement applies only to the reader, and that Janssen himself had already carried out the experiment, very probably at Le Creusot. Consequently, he had one of the keys to the construction of a purpose-built instrument (Fig. 9.2) that would allow him to obtain photographs of the granulation, the best of which would become the reference until the observations made by Bernard Lyot at the Pic du Midi in 1943, 36 years after Janssen's death!

Janssen and Prazmowski's Solar Photographic Refractor

To realize this aim, Janssen therefore turned to an instrument maker who was not just anybody. Before he took refuge in Paris in 1863, Adam Prazmowski was actually a mathematician and astronomer at Warsaw and he had observed the eclipse of 1860

[1] The term *granulation* or *granules* was suggested by Dawes in 1864: Dawes, William, "Results of some recent observations of the solar surface, with remarks", *MNRAS*, **24** (1864), 161–5, p.162.

in Spain. The work that he carried out for Janssen was the subject of a note on "Chemical achromatism" [9], that Janssen presented to the Académie des sciences on 13 July 1874. Prazmowski outlined there how the correction applied to ordinary photographic objectives – where one had "favored the accuracy of the optical image *[i.e., the visual image]* without giving the photographic image the accuracy of which it is capable," and which gave the image "a certain softness in its outline" – did not accord with the desired goals in astronomical photography. It was, therefore, necessary to achieve a correction specific for the photographic domain. In addition, it was not the two extreme colors of the actinic spectrum that should be "brought together," but instead those whose "chemical action has the most powerful effect." Because this action was a function of the material used, Prazmowski explained that "M. Janssen had photographs taken to his specification of the spectra created by prisms made of the flint and crown that would be used in the construction of the objective and the eyepiece – or rather of the equipment that magnified the image created by the objective. The photographs demonstrated that the most powerful action was produced by the portion of the spectrum between the G and H *[Fraunhofer]* lines," that is, between indigo and violet. To achieve and determine the curvatures, he used Biot's results, which led to the two interior surfaces of the lenses in the doublet being in contact, and curvatures that gave a minimum deviation to incident rays that were parallel to the optical axis. There was one last personal touch by Prazmowski, and by no means the least:

> Authors recommend calculating achromatism for the centre, and letting an approximation serve for the edges. I believe that it is useful to follow the opposite course. In the objective built for photographs of the Sun, the marginal rays are rigorously achromatized; achromatism of the centre is approximate, but to such an extent that, in the image, this small error remains imperceptible as a result of the tiny angles of incidence.

As we can see, the investigation was exhaustive, and all these considerations resulted in the realization of the excellent objective 135 mm in aperture, with a focal length of 2 m, constructed for Japan. The judgment of Samuel Langley, the great specialist, as Janssen informed Huggins, was eloquent on this subject: "He admitted to me, with greatest courtesy, that photography with our 5-inch[2] revealed details that, over several years, his large refractor had not revealed to him during half-an-hour" [10].

After he returned from Siam in 1875, Janssen regularly used his refractor from his enclosure at boulevard Ornano. From April 1876 he turned out large-size photographs of the Sun (they were 20–22 cm in diameter), which allowed him to study details of the spots and faculae. In June he declared:

> It is, therefore, nowadays superfluous to emphasize the significance of celestial photography. But it seems to me that this application should now enter a new phase. For the Sun, for example, observatories that are undertaking these studies should start indefinite series, that are as complete as weather permits. These series should consist of very large and very perfect images. [...] The solar images should reproduce the faculae and their exact outlines, spots and their so significant structural details, and finally, the granulations of the surface and their true forms [11].

---

[2] Five French inches. One French inch ("pouce") equalled 27 mm.

Encouraged by the good results from his optics, Janssen set out to carry out an in-depth study of solar photography. The work bore on the spectral sensitivity of emulsions, on improving the fine-grain structure of their layer, and on exposure times.

To avoid over-exposure, he greatly reduced his exposure times, which further reduced the active spectral region, and correspondingly assisted calculation of objectives. But optical perfection was not sufficient. To make the emulsion even more sensitive and as fine-grained as possible it was then necessary to improve the formula for the collodion, and of its major component, gun-cotton, that served to hold the sensitive grains in place. It was Arents who devised an iodide/bromide collodion (where iodide predominated) giving a maximum sensitivity in the violet, and who perfected a procedure that allowed extremely homogenous, sensitive plates to be obtained.

The final difficulty that remained to be solved in obtaining perfect images was reduction of the "photographic irradiation," the diffusion within the sensitive layer that could reach 20 s of arc. Janssen carried out experiments with larger and larger magnifications, giving an image of the Sun with diameters of 20, then 30, and even 50 and 70 cm. In enlarging the size of the solar disk, he correspondingly decreased the flux reaching the plate, but the resolution remained limited by turbulent motion in the atmosphere. It was thus necessary to reduce the exposure time considerably, by developing a better shutter. He therefore conceived a device, which he called a "trappe photographique" *["photographic trap"]*, a very ingenious mechanism, which was able to attain shutter speeds as fast as 1/6,000th of a second! The first such "traps" were made by Prazmowski, then Janssen asked Gautier to make more elaborate ones that did not introduce any vibration.

It is obvious that, for Janssen, photography should in future be considered as a method of discovery in physical astronomy, and he created, without delay, a department for celestial photography in the new observatory of which he was Director.

Granulation and the Non-existent "Janssen's Photospheric Network"

In 1877, he first proved the importance of series by displaying photographs showing the appearance of a sunspot from one day to the next. Then he presented his discoveries concerning the solar granulation and what he called the "photospheric network." Certainly De La Rue and Rutherfurd had already taken fine images showing spots and faculae "but not the details of granulation, which visual instruments have revealed to our eyes" [12]. If he, Janssen, succeeded in capturing "the rice grains," it was because he knew how to ensure he held all the technical trumps in his hand. The shapes of the grains are very varied, but are more or less spherical or spherical in origin. Their material is extremely mobile "particles of solid or liquid matter floating in a gaseous medium" [13]. As for the solar surface, "it is divided into regions of relative calm and activity, which results in the production of the *réseau photosphérique* [photospheric network]":

> Already, examination of photographs embracing just a small number of months shows differences in the constitution of the photospheric network, differences which will teach us about the variations in the forms of solar activity. As such, from now on, solar photography is placed in a position where it can reveal the most important facts about the constitution of the Sun. It is a new method that is opening up before us, whose gains may be linked with those from spectral analysis and the older visual methods to resolve finally, and definitively, the great problems posed by the daytime star [14].

The quality of Janssen's images of granulation is so good that Lockyer requested copies, which Janssen sent him in December 1877:

> Here, finally, is a print on glass of our images that is sufficiently satisfactory. It does, however, still not show all the details on the original. At your request, I intend this proof to go to the Kensington Museum [15].

In addition, Janssen served as an intermediary with Prazmowski who would be able to provide Lockyer with "a chemical objective for solar photography," 5 French inches *[in diameter]* for 1,000 F, or else one of 6 inches for 1,800 F [16]. Janssen would test the objectives at Meudon. Lockyer ordered the 6-inch. Several years later, Janssen would negotiate with Gautier and the brothers Henry for a photoheliograph for Lockyer.

Janssen was convinced that everything was perfect. His optimized optics, on the one hand, and the high speeds of his shutters on the other, left no room for any doubt; what he recorded on this plates was finally objective and permanent proof of the true reality of the phenomena that occurred on the surface of the Sun. He was, therefore, particularly confident of the existence of this "photospheric network," which he believed to have discovered with his images as proof, and of which he never ceased to speak. Sadly, the physicist who was so attentive to the pernicious effects of our atmosphere, did not, however, realize that what photography (which, itself, records everything that it sees) showed him was exclusively caused by inhomogeneities in the *terrestrial* atmosphere: fluctuations in its density and refractive index [17]. He had considered just the dynamic component of turbulence... From 1886, his pupil Stanoïevitch *[Stanojevic]* cast doubt [18] on the existence of the network, without discovering its true cause [19]. Janssen, who obviously did not share that view, nevertheless agreed to play the game, and to present the paper:

> The preceding note that M. Stanoïevitch has asked me to present to the Académie is interesting in this sense: that it calls greater attention to the deformations that the solar atmosphere may produce on images of objects located on the surface of the Sun. But it would be to go too far to attribute to it the production of the photospheric network itself. There are a number of facts, which it is not necessary to discuss here, that conflict with such a conclusion [20].

However, when the images were good, they were breathtaking, and they were, quite rightly, admired by everyone. The photograph of granulation of 10 October 1877 (Fig. 9.5, p. 118), which was widely distributed, with its description as "a photoglyphic proof obtained without any manual intervention," is undoubtedly one of the most beautiful examples. The photograph of the spot of 22 June 1885 (Fig. 9.3), one of the largest observed up to that time, because it measured 2 min of arc (1/16 of the diameter of the solar disk), was also seen worldwide. "A thousand thanks for your

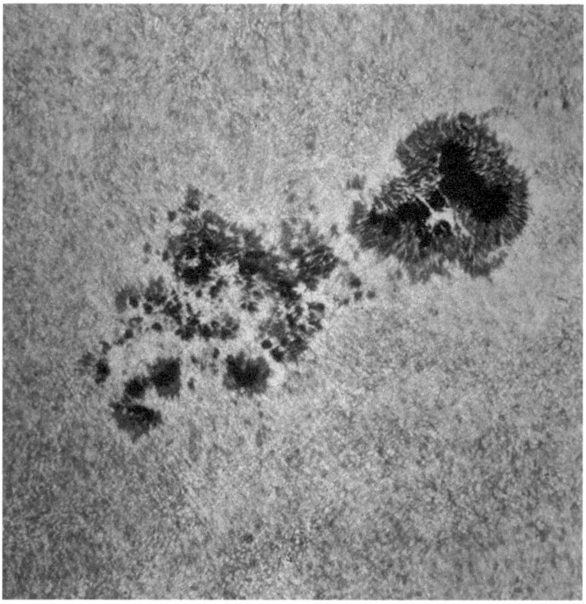

Fig. 9.3 *Sunspots photographed at Meudon on 22 June 1885.* Annales de l'Observatoire d'Astronomie Physique de Paris. © Paris Observatory Library

beautiful photo which is quite the finest I have ever seen" [21], wrote Lockyer, who went even further a few days later: "I am still lost in admiration of your wonderful sun spot" [22]. It was a connoisseur speaking...

The Scientist's Retina

Janssen had, admittedly, been tricked by the Earth's atmosphere which he had spent most of his time observing, characterizing, and eliminating. Against this, the future ahead of scientific photography did not escape him.

Janssen had a sense for phrases, and there is one that the history of photography will not forget: "So, I do not hesitate to say that the photographic plate will soon be the actual retina of the scientist" [23]. It was at the 1882 AFAS meeting in La Rochelle had he said this, and when he recalled it 4 years later, it had undoubtedly already become so widespread that Janssen was convinced that it dated from 1877. He thus claimed, in 1886 [24], that he had mentioned it at the Académie des sciences in December 1877, and that it appeared in the "Notice sur les progrès récents de la physique solaire" *["Note on recent progress in solar physics"]* in the *Annuaire du Bureau des longitudes pour l'an 1879*; where, however, it cannot be found, neither in that note nor in the *Comptes rendus* for 1877. The reference to the eye is hardly surprising, coming from a researcher whose thesis concerned physiological optics. Janssen was particularly familiar with the limits of our vision, which are not those of the "sensitive layer of the photographic plate, by reason of this admirable property *[that it has]* of enabling us to fix images, and to form them with a group of rays that is far more extensive *[spectrally]* than those that affect our retina and, finally, to allow the accumulation of its radiant action over a time that is, so to speak, unlimited" [25].

The Method of Stellar Circles

In 1880, Janssen discovered the "periodicity" *[non-linearity]* of the action of light on photographic materials, and hoped to be able to obtain photographs of the chromosphere by solarization.[3] The same year, with the new approach opened up by the "dry process" and Henry Draper's success, he became interested in the study of nebulae, where it was difficult to obtain images that were "more-or-less complete" [26]. He was then working with a very fast reflector, 50 cm in diameter, and he was insistent that it was absolutely essential to have a photometric control on the photographs. Again a pioneer, he proposed to defocus the images by placing the plate slightly in front of the instrument's focal point. Instead of a black point as the image of a check star, "we thus obtain a circle with a very small diameter, with a nearly uniform tint [...] the degree of opacity of which may be compared with circles with the same origin" [27]. Subsequently, he never ceased advocating his excellent method of "stellar circles" for computing light emitted by stars. It may be noted that Janssen's method was profitably put into practice by Karl Schwarzschild, about 20 years later [28].

In 1881, Janssen created his "photographic photometer," which enabled him to determine the sensitivity curves of emulsions and to compare "the photogenic intensity of plates." This was how he was able to state that "the new gelatin/silver-bromide plates that are being prepared nowadays may be as much as 20 times as sensitive as wet process collodion plates" [29].

The Great Comet of 1881

It was thanks to the use of these new plates, the "extra luminous" 50-cm reflector [30], and to an excellent tracking mechanism for following the movement of the body that "for the first time, in 1881, the *complete* photographic image of a comet has been obtained" [31] (Fig. 9.4). The adjective is important, because Janssen recorded the image in an exposure of just 30 min, while he measured the light from the tail of Tebbutt 1881 III and found that, photographically, it was 300,000 times less than that of the nucleus! Certainly, Draper's photographs were taken a few days earlier, but on them the tail of the comet is hardly visible – at least on those that still exist [32]. Huggins himself had taken only photographs of the spectrum, and he

[3] This is gross overexposure, which, after development, results in reversal of the negative image.

Fig. 9.4 *The comet b 1881, observed at Meudon on 30 January 1881. Photographic print of a drawing from a photograph.* © Paris Observatory Library

asked Janssen for one of his pictures to show a slide at a lecture that he was giving at the Royal Society in January 1882 [33].

At the same time, Janssen discovered that the photograph also showed "in the region occupied by the tail, a number of very small *[faint]* stars, several of which do not appear on any atlas." This observation, more than 17 months earlier than the well-known one by David Gill, seems to have passed completely unnoticed at the time, because Gill's photographs of the comet of November 1882 – and not Janssen's photos of 1881 – were the origin of the Carte du Ciel project! Really, we must believe that, after the helium affair, a lot of people were blinded by the Sun, and by just the Sun, and that they permanently associated Janssen with it…

The Transit of Venus in 1882

Even though he may have been able to bring tears to the eyes of his friend, the composer Charles Gounod, who exclaimed "Oh, how beautiful!", when he explained Kepler's Laws to him, Janssen was never enthusiastic about celestial mechanics.

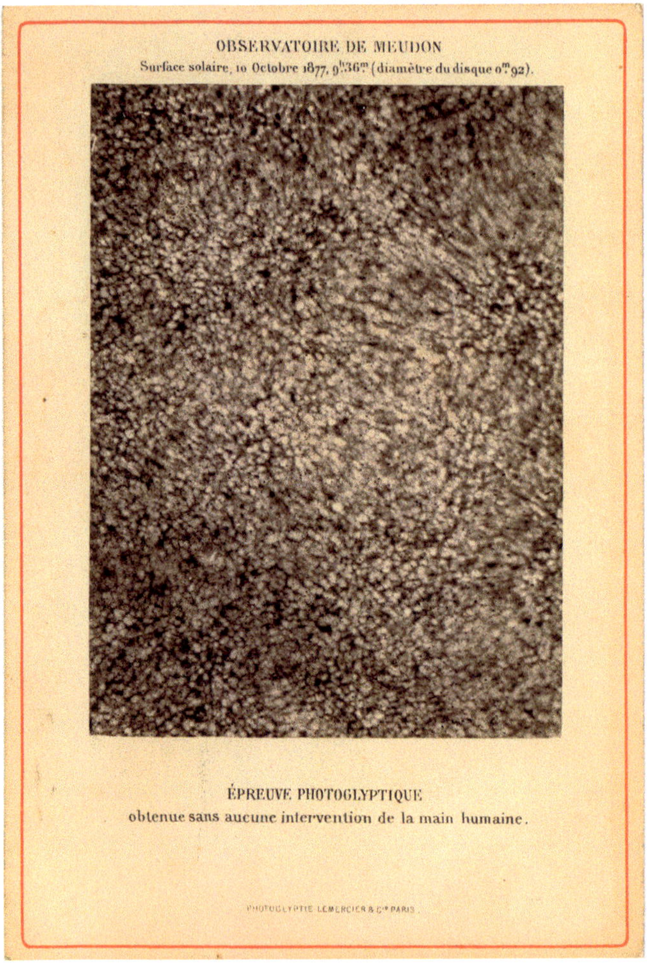

Fig. 9.5 *Solar granulation on 10 October 1877 from Meudon. Photograph by Janssen and Arents.* © Paris Observatory Library

He decided, therefore, not to take part in the observational campaign for the second, nineteenth-century transit of Venus. In December 1882, he simply took the opportunity of the event to carry out physical astronomy:

> The principal aim of these observations was to study a very current question of key significance, both from the point of view of the constitution of the Solar System and that of natural philosophy: I am speaking of the composition of the atmosphere of Venus and of the presence or absence, in that atmosphere, of that aqueous element, which, on Earth, plays such a great role in all the phenomena that are related to the development of life [34].

That was undoubtedly more exciting that a new attempt to determine the solar parallax, and did not require moving to a distant location where all the contacts would be observable. Nevertheless, Janssen did decide to move, to Oran, because in

Algeria the Sun would be higher than in France at the time the planet entered the disk of the Sun. He would make spectroscopic observations and large photographs, where the planet would be 1 cm in diameter. So he left with his photographer, Louis Pasteur, who had replaced Arents at Meudon in 1880, and a serviceman, a former student at the Polytechnique. He transported all the materiel that he had in Japan in 1874, except the revolver, which he did not need.

Because he was rather disappointed by his results (the presence of water vapor was far from being obvious), he extended his stay in the high-plateau region and remained there for about a month, three times as long as he had intended. There, despite excellent observational conditions – the extraordinary "brilliance" of the planet, enabling high-dispersion and high-quality spectroscopes to be used, and the extreme dryness of the site, to his great regret he had to accept the evidence: "I am led to admit that, when one thus eliminates the influence of the terrestrial atmosphere, the optical characteristics of water vapor in the spectrum of Venus are very weak." However, he still tried to retain a glimmer of hope: "This does not mean, to my mind, that this element is absent in Venus ..." [35]

Janssen's *Atlas of Solar Photographs*

Janssen did not fail to use photography whenever he could, during his everyday life at Meudon, with his observational instruments and in his spectral-research laboratory, or during his observational expeditions. "Photography links phenomena across time, just as writing links thoughts and emotions across the ages" [36], he maintained in his speech in 1888 in honor of photography. Whether on the verge of the desert in Algeria, at the edge of the crater of Vesuvius, or at high-altitude observatories, he photographed, or had photographed, everything that could be captured for future generations: stars and their spectra, but also landscapes, clouds, monuments, or instruments. Only portraits of the scientific participants were missing from the roll-call. They exist, but they were reserved for close friends and for that, Janssen went to sit for photographers, such as Nadar and others: "I have, in due course, received your letter and your portrait, which have given us great pleasure. In turn, I would beg you to be kind enough to accept the enclosed, which has just been taken by a photographer who is taking *[photographs of members of]* the Institut, and who asked me to sit for him" [37], he wrote to his friend Huggins in 1879. However, within his family circle, Janssen occasionally liked to play at being a photographer and, in 1881, he did not refuse to take the portrait of some of his cousins, who were delighted with the result.

Janssen's major photographic work is undoubtedly the collection of daily photographs of the surface of the Sun, which he had his photographers obtain. It was Arents for the first 4 years, and then Pasteur, aided by Coroyer, from 1880. In 1903, the collection consisted of 6,000 photographs of the whole Sun, recorded since 1876 on glass plates, 36 cm on a side: an absolute treasure-trove. One hundred years later, just seven remain.

Happily, Janssen published some of them, but that was not as simple as it might appear. Contemporary printers did not know how to reproduce the photographs on ordinary paper. Either engraving had to be used, or actual photographic prints had to be glued into the books. This is what Janssen did in the *Annuaire du Bureau des longitudes pour l'an 1879*, where he published "two photoglyphic proofs of the same region of the solar surface, taken at an interval of 50 min, and showing the rapid transformations of the network of the photospheric granulation" [38]. This solution was satisfactory, but very expensive, especially if one wanted to have large-size proofs. Janssen decided to plunge into this lavish operation, and he did a lot of work with Pasteur to obtain large, high-quality and durable copies. First, he published 12 in the bound edition, a large quarto, of the first volume of the *Annales de l'Observatoire d'Astronomie physique de Paris*, which appeared in 1896. But he wanted to do better and grander things: he selected the 30 pictures that he considered the most significant (and which are not, unfortunately, the best, because of the "photospheric network," which played rather too prominent a part), and in 1903 published the "first part" of the great *Atlas de photographies solaires* [39], where the diameter of the Sun, on sectional plates 44.5 cm wide and 53.5 cm high, is no less than 1.2 m! The work, which weighs some 10 kg was undoubtedly printed in just a few dozen copies, and there was never a second part.

To see anywhere in the literature a photograph of the whole Sun, taken by Janssen, we need to turn to the popular work by Guillemin, *Le Ciel*, which, after its fifth edition of 1877, has a photoglyphic plate of a very fine picture showing spots and faculae, and where, on the original plate, the Sun had a diameter of 204 mm [40].

References

1. Janssen, Jules, "Les méthodes en astronomie physique", *AFAS, Compte rendu de la 11ᵉ session, congrès de La Rochelle (1882)*, 1883, 9–23.
2. Janssen, Jules, "Présentation de quelques spécimens de photographies solaires obtenues avec un appareil construit pour la mission du Japon", *CRAS*, **78** (22 June 1874), 1730–1.
3. Jules Janssen's notebook, BIF Ms 4125.
4. Faye, Hervé, "Communication de l'extrait d'une lettre du P. Secchi concernant des expériences photographiques faites pendant l'éclipse solaire du 8 juillet", *CRAS*, **33** (08 September 1851), 285–7.
5. Faye, Hervé, "Observation photographique de l'éclipse faite avec la grande lunette de M. Porro", *CRAS*, **46** (15 March 1858), 507.
6. Faye, Hervé, "Sur les photographies de l'éclipse du 15 mars présentées par MM. Porro et Quinet", *CRAS*, **46** (12 April 1858), 705–10.
7. Janssen, Jules, "La photographie céleste", in *AFAS, Compte rendu de la 164e session, congrès de Toulouse (1887)*, 1888, 395–409.
8. Janssen, Jules & Follin, Eugène, "Considérations physiologiques sur l'éclairage et applications à l'examen ophtalmoscopique", *Archives générales de médecine*, **18**, (1861), 26–37.
9. Prazmowski, Adam, "Sur l'achromatisme chimique", *CRAS*, **79** (13 July 1874), 107–10.
10. Rough draft of a letter from Jules Janssen to William Huggins, n.d. *[1879]*, BIF Ms 4135-332.
11. Janssen, Jules, "Présentation à l'Académie de photographies solaires de grandes dimensions", *CRAS*, **82** (12 June 1876), 1363–5.

12. *Ibid.*
13. Janssen, Jules, "Sur la constitution de la surface solaire et sur la photographie envisagée comme moyen de découverte en astronomie physique", *CRAS*, **85** (31 December 1877), 1249–55.
14. Janssen, Jules, "Sur la photographie solaire et les faits qu'elle nous révèle touchant la constitution de la photosphère", in *AFAS, Compte rendu du congrès du Havre, août 1877*, 1878, 327–32.
15. Letter from Jules Janssen to Norman Lockyer, 07 December 1877, EUL.
16. Letter from Jules Janssen to Norman Lockyer, 23 December 1881, EUL.
17. Rösch, Jean, "Photographie de la photosphère et des taches solaires, *l'Astronomie*, **71** (1957), 129–41.
18. Milogradov-Turin, Jelena, "Origin of the photospheric network – The main astrophysical research of Dorde Stanojevic", *Publ. Astron. Obs. Belgrade*, **80** (2006), 265–8.
19. Stanoïevitch, Georg, "Sur l'origine du réseau photosphérique solaire", *CRAS*, **102** (12 April 1886), 853–6.
20. Janssen, Jules, "Observations relatives à une communication de M. G.-M. Stanoïewitch sur l'origine du réseau photosphérique solaire", *CRAS*, **102** (12 April 1886), 657.
21. Letter from Norman Lockyer to Jules Janssen, 10 March 1886, BIF Ms 4136-370.
22. Letter from Norman Lockyer to Jules Janssen, 24 March 1886, BIF Ms 4136-371.
23. Janssen, Jules, "Les méthodes en astronomie physique", *AFAS, Compte rendu de la 11e session, congrès de La Rochelle (1882)*, 1883, 9—23.
24. Janssen, Jules, "Note sur la constitution des taches solaires et sur la photographie envisagée comme un instrument de découverte en astronomie", *CRAS*, **102** (11 January 1886), 80–2.
25. *Ibid.*
26. Janssen, Jules, "Sur les photographies de nébuleuses", *CRAS*, **91** (02 November 1880), 713-4.
27. Janssen, Jules, "Sur les photographies de nébuleuses", *CRAS*, **92** (07 February 1881), 261-5.
28. Hearnshaw, John, "Photographic stellar photometry in the late 19th and early 20th centuries", in *100 Years of Observational Astronomy and Astrophysics*, C. Sterken and J. Hearnshaw, eds, 2001, 171–83.
29. Janssen, Jules, "Sur la photométrie photographique et son application à l'étude des pouvoirs rayonnants comparés du Soleil et des étoiles", *CRAS*, **92** (04 April 1881), 821–5.
30. "La grande comète B 1881 photographiée à l'Observatoire de Meudon", *La Nature*, No. 441, 12 November 1881, 383–4.
31. Janssen, Jules, "Note sur la photographie de la comète b 1881 obtenue à l'Observatoire de Meudon", *Annuaire du BdL pour 1882*, 773–86.
32. Pasachoff, Jay M., Olson, Roberta J.M., and Hazen, Martha L., "The earliest comet photographs: Usherwood, Bond and Donati 1858", *Journal for the History of Astronomy*, **27** (1996), 129–45.
33. Letter from William Huggins to Jules Janssen, 13 December 1881, BIF Ms 4135-324.
34. Janssen, Jules, "Note sur l'observation du passage de la planète Vénus sur le Soleil", *CRAS*, **96** (29 January 1883), 288–92.
35. *Ibid.*
36. Janssen, Jules, "En l'honneur de la photographie", *Bulletin de la Société française de photographie*, **4** (1888), 166–8.
37. Letter from Jules Janssen to William Huggins, n.d. *[1879]*, BIF Ms 4135-332.
38. Janssen, Jules, "Progrès récents de la physique solaire", in *Annuaire du BdL pour l'an 1879*, 623–85.
39. Janssen, Jules (assisted by M. Pasteur), *Atlas de photographies solaires*, Gauthier-Villars, Paris, 1903.
40. Guillemin, Amédée, *Le Ciel*, Hachette, Paris, 1877.

Chapter 10
From Caroline Island to Washington

> *"Tout effort, quelque obscur qu'il soit, devient grand quand il a un mobile élevé et généreux".*
>
> *[Any effort, however obscure it may be, becomes great when it has a noble and generous motive.]*
>
> Jules Janssen [1]

Westward Ho! – for the Antilles

In 1883, we find Janssen leaving again for a new eclipse, when he traveled far across the Pacific to Caroline Island. "It is an island a league wide that is so flat that a hillock 1.5 m high is a noticeable rise" [2], was what he was told by the sailors on the *Saint-Nazaire* when he boarded it on 5 March.

It was the first time that Janssen had left on an expedition under such favorable conditions. He was accompanied, not just by the photographer Pasteur and an assistant from the observatory, but also by an astronomer he had recently recruited at Meudon. This was Léopold Trouvelot, whom Flammarion had introduced to him on 27 April 1882: "M. Trouvelot, an amateur astronomer who carries out spectroscopic studies, wishes to be introduced to you" [3]. Having come back from the United States after staying there for some 10 years (it was political events that caused him to leave France), Trouvelot was, in fact, trying to capitalize on his immense talent as a draughtsman, and he probably talked about spectroscopy to attract Janssen's attention. After having devoted himself to raising a potential substitute for silkworms, gipsy moths (in fact a voracious species that escaped and caused extensive, and still continuing, defoliation of trees and shrubs in a major ecological disaster), Trouvelot had become generally known for his extremely numerous drawings of the Sun and the planets that he made while in the United States, during several months at Harvard

F. Launay, *The Astronomer Jules Janssen: A Globetrotter of Celestial Physics*, Astrophysics and Space Science Library 380, DOI 10.1007/978-1-4614-0697-6_10,

Observatory and over several years from his private observatory in Cambridge [4]. The high point was the publication in New York, in 1882, of a portfolio of 15, large-sized chromolithographs reproduced from original pastels that were works of art of indisputable documentary and artistic quality [5]. Trouvelot had also had the chance to meet the Harvard astronomers in Wyoming, where he had gone to observe the eclipse of 1878, and he had the honor of seeing his report printed in a US Naval Observatory publication. Janssen therefore had three collaborators with him, of whom he was, administratively, the leader. But that was not all: two very eminent foreigners decided to accompany him. These were Johann Palisa, from the Vienna Observatory, a major discoverer of asteroids, and Pietro Tacchini, Director of the Rome Observatory, a great specialist in the Sun, whom Janssen knew well, and with whom he had been on best terms for many years. Janssen was delighted with his team: "The personnel on the expedition are pleasant and show a respect for me. I find their society very agreeable" [6], he wrote to Henriette during the course of the 13-day crossing, which first took him to Guadeloupe. That certainly helped him to forget the extremely painful period of perpetual rolling and pitching during the first week of the voyage. At Pointe-à-Pitre, the team came to a magnificent roadstead, where they disembarked:

> We landed and went for a walk in the town, and particularly in the market. The black element is dominant here. All you see are great devils of negresses, done up in provocative madras cotton headscarves and wrapped in robes with the waistline under the armpits, with endless skirts, large enough to hide a regiment, who stride around the town with an insufferable gait and swaying of the hips.
>
> Universal suffrage has put the colony into the hands of the blacks. It is they who nominate the deputy, who form the municipal council and who, all in all, have all the power. What a reversal of things! The white element is furious, humiliated, and sickened. That hardly turns them into republicans [7].

At the roadstead at Fort-de-France in Martinique, Janssen witnessed the procedure of coaling ship, which gave him the opportunity of a penning a passage with particularly vivid imagery:

> In this country, which has so little progressed since the beginning of the century, they still do not know about using machines to carry out coaling. The ship ties up to the quay and they set out two large gangways to connect it to the shore. Then you see a column of black women appear, black by nature, and black by toil. Each carries a basket full of coal on her head. One hand steadies the basket, the other rests on the hip. Their clothes were probably originally colored. [...] They have to [...] step on a small rocker that counts the number of baskets. It is calculated that it requires 12–15,000 baskets to supply the ship. [...] The procession of porters goes on and on, and darkness begins to fall. A black man arrives on the quay and starts to tap on a tambourine: a rhythmic, monotonous noise, which he accompanies by a light whistling, which rise above the dark notes of the bass. With this national chant, the women come to life and increase their pace; they sway their hips and straighten up beneath their loads, which seem to become lighter. These porters who, just a moment ago, seemed to be miserable and hideous creatures, gain in character; their silhouettes, bathed in the twilight, take on classical poses and are strikingly outlined against the dying light of the day [8].

The Isthmus of Panama and the Marquesas

Arriving at Colon, Janssen saw Charles de Lesseps (the son of Ferdinand), who showed him, together with Palisa and Trouvelot, the construction work going ahead for the Panama Canal.[1] "De Lesseps is here in the middle of a court. The millions that have been thrown away here have created a town, but a town that would provide few candidates for the prize that the Académie Française awards for "vertu"[2] *[the Prix Montyon]*" [9]. Janssen, who had to take care of the crossing of the isthmus, made use of Lesseps in dealing with the managers of the Panama Railway, that "terrible railway," before treating the board members to champagne. Really, Janssen was on top form!

He was, however, very annoyed to learn from the commander of the *L'Éclaireur*, specially sent from Panama to meet him, that he intended to make part of the crossing under sail. *L'Éclaireur* was a naval vessel put at the expedition's disposal by the ministre de la Marine, and Janssen found that part of the programme "absolutely unacceptable." As a result, he asked the commander to buy an additional 100 tonnes of coal to cover the whole 1,200 leagues under steam. We know how persuasive Janssen could be: "Faced with my clear, firm declaration, made in the name of science, he told me that he would carry out my wishes" [10].

Everything went as Janssen hoped, and on 19 April he was in the Marquesas: "We are going to land and see these fierce Maoris of the Marquesas, who are very handsome, very strong, very warlike, and very cannibalistic. But we have tamed them, and there is nothing to fear, even in an excursion into the interior" [11]. Henriette was doubtlessly delighted to hear it.

The Eclipse of 9 May 1883 at Caroline Island

Luck was again with Janssen, and he had not made the long and grueling trip (transporting a considerable number of instruments) for nothing:

> We have had yet another success. I have definitely proved the existence of the Fraunhofer spectrum [perhaps some one hundred lines] in the Corona, which demonstrates the existence of cosmic material in the region around the Sun. This is a fine success, but it took a lot of trouble to get it. This trip has been the most exhausting of all those that I have made – infinitely so. I have had to concern myself with everything and everyone on this desert island, with no mechanic, almost without help, and will all the responsibility. I have been working in an outrageous manner. On the eve of the eclipse, I spent from 9 o'clock to 2 in the morning in

[1] This was the first attempt to construct a canal across central America at the beginning of the 1880s. The work was interrupted in 1889 after the bankruptcy of the company directed by the Frenchman, Ferdinand de Lesseps, a bankruptcy that created an enormous financial scandal. The work was restarted in 1904 by an American company, and the canal was finally opened in 1914.

[2] This refers in particular to the award for the most charitable and/or courageous act by a poor Frenchman.

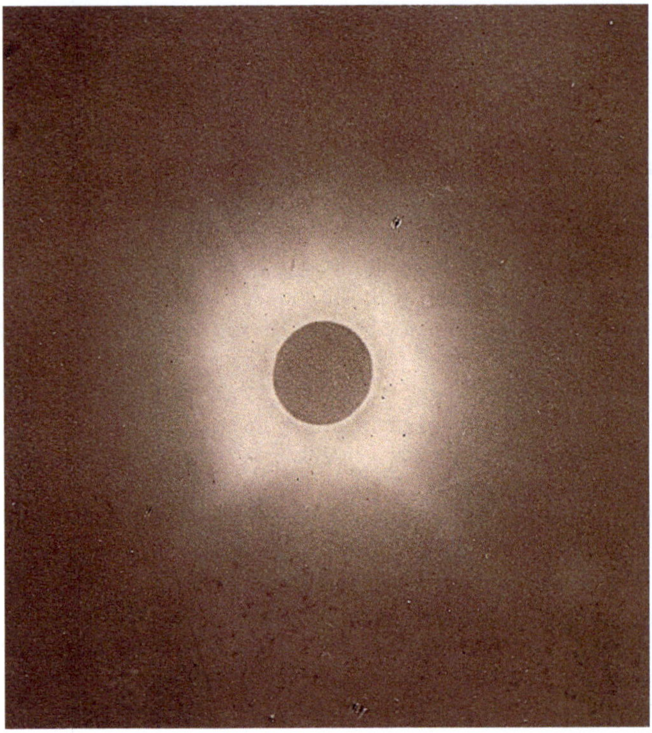

Fig. 10.1 *Photograph of the corona during the total eclipse of 6 May 1883.* Annuaire du BdL pour l'an 1884. © Author's collection

> the most exhausting positions, adjusting my instrument. On the morning of the eclipse, we had rain and stormy weather. All these gentlemen thought that the expedition was lost. Finally the sky cleared 10 min beforehand, and throughout the phenomenon, we had magnificent weather. Everything went well, everyone carried out their task well, and after the event, our pleasure was immense [12].

Indeed, Pasteur, who had eight cameras, obtained superb pictures of the corona, and a genuine print on paper of one of them was to be included in the report published in the *Annuaire du Bureau des longitudes* [13] (Fig. 10.1). It was taken with an objective 21 cm in diameter, with a focal length of 1.20 m, and an exposure time of 5 min. The corona is more extensive than could be seen visually, but the phenomenon was limited *[in extent]* and did not vary during totality. On the other hand, Janssen tried, for the first time, to make a photometric measurement of the luminous intensity of the corona, which was greater than that of the Full Moon, by photography. Also for the first time, he observed the "Respighi rings," colored images of the eclipsed Sun that were obtained by placing a prism in front of the refractor's objective. Immediately after the eclipse, the four astronomers wrote a brief report, which was read to the others and countersigned by them. Trouvelot made a drawing of the corona. Like Palisa, with whom he had shared the task, he had searched, out to a distance of 15° from the center of the Sun, for the hypothetical intramercurian planet that could perturb the orbit of Mercury, and which the doctor Lescarbault had thought he observed in 1859, and whose orbit Le Verrier had calculated. Naturally, the photographs did not show anything either.[3] Tacchini, for his part, had studied the polarization of the corona and carried out spectroscopy. He found an obvious resemblance between "the constitution of some parts of the corona and the spectrum of comets" [14].

In addition, despite what he said, Janssen had 17 men from the *L'Éclaireur* at his disposal, and the effectiveness of the sailors was obviously commensurate with expectations. Moreover, relations with the American expedition, directed by Holden, were excellent – in particular, the Americans lent one of their "chalets" – to such an extent that the day after the eclipse, Janssen invited his colleagues to dinner. "You would have thought that we were in Paris. Great cordiality" [15], he wrote to Henriette, to whom he also added that his health remained remarkably fine.

Tahiti, the Sandwich Islands, and America

The return journey took Janssen to Tahiti, where "the Expedition was welcomed by the chiefs according to ancient custom." Unfortunately, Janssen could not attend the last celebrations, because he had an acute abscess on his arm. He had completely recovered when, on route for San Francisco, he passed the Sandwich Islands (Hawaii). He asked that they put in there because he wanted to stay for a week to carry out some geological research on an eruption that had just occurred. It is there that the most remarkable volcanoes "in the world"[4] [16] occur at an altitude of 4,300 m. He went to the crater of Kilauea, which was how "a night passed in the great crater, the most remarkable in the world, and on the edge of a molten lake, enabled *[him]* to make a study that resulted in *[finding]* strange analogies between these volcanic phenomena and those of the surface of the Sun" [17].

The group returned to *L'Éclaireur*, which carried them to San Francisco, where the expedition stayed from 9 to 21 July, and was thus able to take part in the 14th-July[5] celebrations by the French colony. Janssen and Trouvelot also wanted to visit the observatory on Mount Hamilton (the Lick Observatory), an excursion that required several days, before crossing America to see the observatories at Madison, Chicago (Yerkes), Cambridge, and finally, Washington. There, Janssen met his famous friend Alexander Graham Bell, who made "his stay in that beautiful city,

[3] It was, in fact, general relativity that explained the observed irregularities in the motion of Mercury.

[4] In English in the text.

[5] Bastille Day (the French national holiday)

extremely pleasant and extremely fruitful" [18]. It was from New York that the packet-boat *Canada* finally brought everyone back to France on 15 August.

On 14 September, Huggins congratulated Janssen in the carefully chosen terms that francophone British people knew how to use so well: "Cher conquérant du Soleil, je vous félicite de vos nouvelles conquêtes sur le terrain solaire et ses colonies coronales" *["Dear Conqueror of the Sun, I congratulate you on your new conquests of the solar terrain and its coronal colonies."]*

Planet Henrietta

Janssen did not receive the letter immediately, because he was attending a meeting in Austria. Before he left, he was lucky enough to meet, in the Paris railway waiting room, his colleague Maurice Lœwy, the future Director of the Paris Observatory, who had lived in Vienna. The latter spared him from having to be concerned about the German language, and he had the pleasure of visiting Salzburg with him. He was warmly received in Vienna: "the Director of the Observatory *[Weiss]* was pleased, and Palisa was cheerful" [19]. On the Sunday, the meeting left on a trip to Mount Kahlenberg, before attending a dinner given by the government. Toasts followed, and Janssen was, of course, asked for a speech. As his subject he took "scientists' wives." Janssen first congratulated the Vienna ladies, and especially Mme Weiss, by thanking her for "the welcome extended the day before in a Palace where one often watched the Heavens" [20], and then he let his emotions have free flow:

> I examined the role of a wife as a companion and often collaborator of scientists. I spoke of wives, who fulfilled this role historically. I showed how this union, which is already so full of sweetness, took on a supreme charm when the wife's mind was able to follow that of the man she loved into the higher regions that the science of the heavens opened to human intelligence: how this contemplation of the beauties of the intellectual world by two, united souls had the power and strength to unite them even more closely. As for myself, I then added, if I am allowed to mention myself in all humility, and to add myself, with all due modesty, to the list of illustrious examples that I have just quoted, I would say that I have experienced this delight, and that here I must express my gratitude to my dear companion, to the one who has followed me, helped me, and even advised me in my travels for science! – Applause – Everyone rose and came to clink glasses with me, drinking to Madame Janssen. Whilst I was talking with infectious emotion of the happiness of two souls, who after having fallen in love on Earth, slowly soar towards Heaven, there to enjoy the delights of an intellectual, refined love, I saw silent tears all around me. [...] When the audience had calmed down, Palisa rose and said that in recognition of what M. Janssen, his mentor and head of the expedition *[represented to him]*, he asked me to be godfather to one of his planets – Applause. I said that I was deeply touched by the honor that M. Palisa accorded me, and that, in truth, his company had been so agreeable to me that the word "recognition" could not be used, and I asked that of "friendship" should simply be substituted, and that because I was asked for a name for the body, the discovery of which was thanks to M. Palisa's talents, I could do no better than to make the deed suit the words, and to give this planet the name of the person, whom I had just introduced to the gathering and regarding whom I had expressed my grateful feelings; as a consequence, I begged the gathering to accept the name of Henrietta, by which Madame Janssen was baptized. Everyone cheered your name, and everyone returned to congratulate me or rather, to beg me to pass on their congratulations to you.

So the name of my life's dear companion is now thus inscribed in astronomical ephemerides, and will live longer than those of many princesses.

I am delighted by this and, in congratulating you, kiss you and Antoinette tenderly – if an immortal will be kind enough to allow it [21].

How Henriette's heart must have pounded on receiving this news!

A Dwelling at Père-Lachaise

Now that his wife had a place in the heavens, it was time for Janssen to concern himself with finding a last resting place for her on Earth, but which might as well be one that was well located. He dealt with this in September 1884, a few days before leaving again for Washington for the famous meridian conference. After a long meeting with the Minister for Foreign Affairs at which he received his letter of accreditation, he went on to the Père-Lachaise Cemetery, where he arranged things:

I arrived at the cemetery and the matter was arranged with Henri *[Le Roux]*. We have 3 m of ground, 1 m frontage, very close to Arago, that is to say on the boulevard des Italiens at Père la Chaise. – It is a dream to be located there. At this point there was a pathway, which Henri will remove to give us a place. I paid on the spot – 1,045 F. So, we have, my dear friend, a place for ourselves and ours to rest, and we shall be in the most cheerful and busiest spot of the Père. – If there is any rumor of the sale of plots at Montmartre, arrange for an exhumation (you have the date; there are two coffins, a regular*[–sized]* one, and a small one of Father's) and have those two coffins interred in our present plot, on a temporary basis. So we are now all set [22].

It is indeed there that Janssen rests with his wife, his daughter, and his parents, transferred from the cimetière du Nord (Fig. 10.2). The tombstone carries the fine profile from the Janssen Medal. (Janssen established several prizes: at the Académie des sciences, the Société astronomique de France, the Société de géographie, the Société de photographie …). It is a replica of the engraving made by Alphée Dubois in 1872 for the Janssen—Lockyer Medal.

The Washington Conference of 1884

Almost in the same breath, Janssen continued:

I have just seen the Minister for Public Education, with whom I discussed the assignment – the allowance is set at 150 F per day – with that I could be a representative. It is the allowance of plenipotentiaries.

So Janssen was almost a minister … and almost rich: he immediately went to purchase a soft hat! Indeed, his mission was of significance, because he was to represent France at the conference that would adopt a "prime meridian," that is to say the meridian that would be the origin for the measurement of longitude, and also,

Fig. 10.2 *Janssen's tomb at Père-Lachaise cemetery in Paris.* © Françoise Launay

although it was not in the initial programme,[6] eventually attempt to have the decimal system applied to the measurement of angles and of time. The Conference was to take place in October, so Janssen embarked at Le Havre on 13 September. The crossing lasted 11 days and Janssen was again treated to four small storms, which did not prevent him from reading a work on Mme de Lamartine, the moving résumé of which that he sent to Henriette ended with these words: "For me, the song of the poet is all the more beautiful, and above all more moving, when I knew the bitterness that saddened the end of that life, than if the man had always been happy" [23].

[6] The title of the Conference was actually: "International Conference for the purpose of fixing a prime meridian and a universal day".

Fig. 10.3 *Victor Hugo and Jules Janssen. A slight family resemblance…* Félix Potin album. © Author's collection

As soon as he arrived in New York, he received a visit from a reporter from the *Evening Telegram*, which, that very evening, published an article entitled "Arrival of M. Janssen" [24]. which gives an idea of the reputation of the French representative, and of the curiosity that surrounded him; but Janssen suffered, because the journalist did not speak a word of French, and his dear interpreter, Henriette, was not there. The next day it was the turn of the *New York Herald*, whose journalist published some 60 lines enumerating all *[Janssen's]* titles and publications, and reported that Janssen was "A low sized, well built elderly gentleman, with long snow white hair, and a face not unlike Victor Hugo's" [25] (Fig. 10.3). Janssen invited Graham Bell to dinner, who came to see him with his father, and then those two gentlemen invited Janssen to a ballet performance: "We decided that in view of my great knowledge of English, it was the best thing that they could offer me – things must be proportionate!" He visited a school; saw Barnard, the Director at Columbia College, an old, very deaf scientist, but one full of serenity, whom he regretted had been excluded from the Conference in favor of some naval officer; and was invited to dinner by a banker, whose wife talked to him about … Lamartine!

The Battle Over the Meridian

At the Conference, France had just two representatives: Albert Lefaivre, plenipotentiary minister and consul general, who was to support Janssen in an exemplary fashion, and Janssen. They were charged with promoting a neutral meridian, which surprised the English-speaking delegates. The President was the American, Admiral Rodgers. The three Secretaries were General Strachey, a British delegate, Janssen, and Dr Cruls, the Brazilian delegate.

After quite a number of preliminary discussions and votes dealing with the organization and proceedings of the conference, the American delegate Lewis Rutherfurd proposed Greenwich as the prime meridian, as recommended by the Seventh International Geodesic Conference held in Rome in 1883[7] [26]. But Janssen demanded that a ruling should be obtained about the very establishment of a prime meridian. This proposal being adopted unanimously, it remained for the principles for the choice of the meridian to be defined, and so battle was joined:

> During the course of this week, we have had a truly terrible session. We proposed a neutral meridian and I made a major speech [27] to explain all its geographical, and above all moral, advantages, by showing through the history of old meridians that it was always through national pride that attempts at unification of Longitudes have failed. The English and the Americans, who expected to see us defend Paris[8] [28], a subject on which they counted on being able to defeat us easily, seeing the strength that we gained from such a proposition that was significant both in concept and in being disinterested, were furious, and an attack against me, or rather against my proposition, started. They went for me, one after the other, trying to demolish my meridian by all possible arguments. After my main speech I had to make 7 or 8 others to reply to each person on the grounds that he had chosen. [...] I fought for 4 h tirelessly. When I left this boiler-room, my shirt was wringing wet: it took two days to dry. They had planned to attack me one after the other, each on different grounds. I was alone against forty, among whom were the principal English[9] and American[10] scientists, but the struggle aroused me so much that ideas and expressions came effortlessly. Besides, almost everyone congratulated me, and were kind enough (especially the gallery) to say that the session had been magnificent and that they had been enchanted to see a scientist who was an orator. The moral success belonged to France. This was all we could hope for, because everyone had precise, formal instructions to vote for Greenwich. But our attitude needed to be a noble one, and it was [29].

The task had been greatly complicated by the fact that Janssen's opponents all spoke in English and that, as Janssen explained "it was based on the notes that

---

[7] This meeting was less diplomatic in character than the Washington Conference, whose resolutions would actually represent a political act.

[8] As the French delegates did, although feebly, the year before at Rome.

[9] Captain Sir F.J.O. Evans, Royal Navy; Professor J.C. Adams, Director of the Cambridge Observatory; Lieutenant-General Strachey, Member of the Council of India; and Sandford Fleming, representing the Dominion of Canada.

[10] Rear-Admiral C.R.P. Rodgers, U.S. Navy; Lewis M. Rutherfurd; W.F. Allen, Secretary, Railway Time Convention; Commander W.T. Sampson, U.S. Navy; Professor Cleveland Abbe, U.S. Signal Office.

M. Lefaivre hurriedly gave me that I had to form an idea of their arguments and to which I had to reply immediately" [30]. It may be noted that Lefaivre, a former student of the École normale supérieure, and a great historian, was a former Siege aeronaut, which would not have harmed his close relations with Janssen ... As only France, Brazil, and the Dominican Republic voted for a neutral meridian[11]:

> *Resolved:* That the initial meridian should have a character of absolute neutrality. It should be chosen exclusively so as to secure to science and to international commerce all possible advantages, and especially should cut no great continent —neither Europe nor America.
>
> Twenty-one noes and three ayes. [...] The resolution is, therefore, lost [31].

there was no point in continuing the battle, and France took no part in "the discussion of the choice of the national meridian destined to become the international meridian" [32], during which all the advantages of Greenwich were developed. When it came to the vote for Greenwich, France and Brazil abstained. Only the Dominican Republic voted against.

Janssen spent the whole of the following week drafting the protocol of the Conference, a 216-page volume in the French version [33], and he had to reconstruct from memory the text of his contributions, because there was no means of finding French-speaking stenographers. The working conditions were terrible: "And because here everything is done with breathtaking rapidity, I had to spend two days without getting undressed. [...] Added to that, the whole of the American press is hot on our heels every day. You can imagine whether it is a bed of roses" [34].

The Non-existent Battle Over the "Metric System"

Before the vote, the Spanish delegate spoke to say that, in voting for Greenwich, his government expressed the hope that Britain and the United States would accept the French system of weights and measures. A split appeared, and Janssen plunged in, without any real conviction to start with:

> While the vote *[in favor of Greenwich]* had taken place as arranged and expected, I did gain the sympathy and respect of everyone. And I gathered the fruits of this at the time when a vote was proposed in favor of continuing studies about the application of the decimal system to the measurement of angles and that of time.[12] Basically, the meeting wanted nothing to do with it. The situation appeared so bad that I sent a coded telegraph message via M. Lefaivre to the Minister, asking if we should persist in demanding it. We were told to persist, unless complete failure was certain [35].

So Janssen plunged in and, deferring to him, the President agreed to ask the meeting if it wished to discuss the subject. Only the British and the Americans voted against.

---

[11] This vote is not mentioned in *Greenwich time: and the discovery of the longitude* by Derek Howse, who gives only the main adopted resolutions (with a numbering of his own), in his analysis of voting country by country (Table II, pp. 146–7).

[12] It is easy to see that this was not at all about the adoption of the metric system, as is very often stated to be the case.

> Then, I spoke again, and when it came to the final vote, the English and the Americans, seeing that they were beaten, voted for *[the resolution]*. This was a real success, and shows utterly how if, on the question of the Meridian, everything had not been agreed in advance, I would perhaps have swung the vote. Will all this effort be appreciated in France? Certainly not, but the feeling of having fully served one's country should suffice, and it does suffice for me [36].

Janssen would have liked to know whether after all this, he was mentioned in the French newspapers … And, as he was unable to sail until the 12 November, instead of the 5th, he took advantage of this to make an excursion to Canada, perhaps on the advice of Lefaivre, who had lived there for several years as French consul.

Janssen was, in the end, rather satisfied by the Conference, the work of which had lasted the whole month of October. In the report that he made about it for the Académie des sciences, he gave the following appreciation:

> This result is considerable. But its significance lies mainly rather with the principles that the Conference enunciated, rather than with the solutions that it adopted.

The establishment of a unique meridian and universal time; the unification of the astronomical and civil days[13]; and the extension of the decimal system are reforms that the progress of Science and international relations have rendered both opportune and desirable.

But, in the application of the principles, the Conference has been less gratifying. In choosing the prime meridian, it allowed itself to be too easily seduced by the practical and immediate advantages offered by a meridian that was already widespread, and it mistook the conditions that would have assured that its work received universal and definitive adoption.

As far as we are concerned, we have, on this question, upheld the role dictated by our past, our traditions, and the very character of our national spirit [37].

References

1. Letter from Jules to Henriette Janssen, 30 March 1883, BIF Ms 4133–261.
2. Letter from Jules to Henriette Janssen, March 1883, BIF Ms 4133–255.
3. Letter from Camille Flammarion to Jules Janssen, 27 April 1882, BIF Ms 4135–165.
4. Herman, Jan K. and Corbin, Brenda G., "Trouvelot: From Moths to Mars", *Sky & Telescope,* Dec. 1986, 566–8.
5. Trouvelot, E.L., *The Trouvelot astronomical drawings, with Manual*, New York, Charles Scribner's Sons, 1882.
6. Letter from Jules to Henriette Janssen, March 1883, BIF Ms 4133–255.
7. Letter from Jules to Henriette Janssen, 19 March 1883, BIF Ms 4133–256.
8. Draft of a letter from Jules to Henriette Janssen, 21 March 1883, BIF Ms 4133–258.
9. Letter from Jules to Henriette Janssen, 28 March 1883, BIF Ms 4133–257.
10. *Ibid.*

[13] Until then the astronomical day started at midday, whereas the civil day started at midnight, 12 hours earlier.

11. Letter from Jules to Henriette Janssen, 19 April 1883, BIF Ms 4133–263.
12. Letter from Jules to Henriette Janssen, 09 May 1883, BIF Ms 4133–265.
13. Janssen, Jules, “Mission en Océanie pour l’observation de l’éclipse totale de Soleil du 6 Mai 1883”, *Annuaire du BdL pour l’an 1884*, 847–77.
14. *Ibid.*
15. Letter from Jules to Henriette Janssen, 09 May 1883, BIF Ms 4133–265.
16. Letter from Jules to Henriette Janssen, 21 July 1883, BIF Ms 4133–270.
17. Janssen, Jules, “Mission en Océanie pour l’observation de l’éclipse totale de Soleil du 6 Mai 1883”, *Annuaire du BdL pour l’an 1884*, 847–77.
18. *Ibid.*
19. Letter from Jules to Henriette Janssen, 16 September 1883, BIF Ms 4133–271.
20. Letter from Jules to Henriette Janssen, 18 September 1883, BIF Ms 4133–272.
21. *Ibid.*
22. Letter from Jules to Henriette Janssen, 09 September 1884, BIF Ms 4133–273.
23. Letter from Jules to Henriette Janssen, 26 September 1884, BIF Ms 4133–276.
24. Letter from Jules to Henriette Janssen, 27 September 1884, BIF Ms 4133–277.
25. *Ibid.*
26. *Septième Conférence géodésique internationale tenue à Rome en octobre 1883, Rapport et résolutions*, Rome, Imprimerie royale, 1883.
27. Janssen, Jules, “Rapport sur le congrès de Washington et sur les propositions qui y ont été adoptées, touchant le premier méridien, l’heure universelle et l’extension du système décimal à la mesure des angles et à celle du temps”, *CRAS*, **100** (09 March 1885), 706–26.
28. Gapaillard, Jacques, *Une histoire de l’heure en France*, Paris, Vuibert, (in press, 2011)
29. Letter from Jules to Henriette Janssen, 13 October 1884, BIF Ms 4133–278.
30. *Ibid.*
31. *Conférence internationale tenue à Washington pour l’adoption d’un premier méridien unique et d’une heure universelle, octobre 1884*, Procès-verbaux des séances, Washington, Gibson Bros., 1884. [The English-language version is: *International Conference held at Washington for the Purpose of Fixing a Prime Meridian and a Universal Day; October 1884*; Protocols of the Proceedings; Washington, D.C., Gibson Bros, Printers and Bookbinders, 1884], p. 84–85.
32. Janssen, Jules, “Rapport sur le congrès de Washington et sur les propositions qui y ont été adoptées, touchant le premier méridien, l’heure universelle et l’extension du système décimal à la mesure des angles et à celle du temps”, *CRAS*, **100** (09 March 1885), 706–26.
33. *International Conference held at Washington for the Purpose of Fixing a Prime Meridian and a Universal Day; October 1884*; Protocols of the Proceedings; Washington, D.C., Gibson Bros, Printers and Bookbinders, 1884.
34. Letter from Jules to Henriette Janssen, 13 October 1884, BIF Ms 4133–278.
35. Letter from Jules to Henriette Janssen, 27 October 1884, BIF Ms 4133–279.
36. *Ibid.*
37. Janssen, Jules, “Rapport sur le congrès de Washington et sur les propositions qui y ont été adoptées, touchant le premier méridien, l’heure universelle et l’extension du système décimal à la mesure des angles et à celle du temps”, *CRAS*, **100** (09 March 1885), 706–26.

Chapter 11
Janssen and Edison's Phonograph

> *"Le problème de reproduire artificiellement la voix humaine est un des plus étonnants de ceux que l'homme ait pu se proposer."*
>
> *[The problem of artificially reproducing the human voice is one of the most amazing of those that Man could ever set himself.]*
>
> Jules Janssen [1]

The Meeting at Bath

"Tomorrow, I shall go to see the famous phonograph" [2] Janssen wrote to his wife on 5 September 1888. Janssen was in England, at Bath, where the 58th meeting of the BAAS was being held. Janssen was diligent in attending these meetings. An inventor himself, and always concerned to see science and techniques advanced, he was naturally very interested in the inventions of others. Edison's phonograph was not completely new in 1888 because an early model had been presented to the Académie des sciences in 1878, but it was a different version that had just arrived in Europe.

The next day, the news was more precise: "I also send a kiss to Antoinette, who will receive my next *[letter]*, in which I shall tell her about an autograph phonograph[1] that the great Edison is sending me personally" [3]. Antoinette's promised letter left Bath on 11 September. Janssen first, quite rightly, gave his daughter details of the great success of the communication of his research in progress on the spectrum of oxygen, which had led him to look into the deeper problems of molecular mechanics:

> *[Sir William]* Thomson *[Lord Kelvin]* has been so enthusiastic in seeing all the horizons that the density-squared law[2] opens up in spectral analysis and in molecular mechanics that he has said that it was a splendid discovery and the whole room applauded his speech.

---

[1] In English in the text.

[2] Janssen had actually determined experimentally that the law governing the appearance of molecular bands was a function of the product of the depth of the gas times the square of its density.

F. Launay, *The Astronomer Jules Janssen: A Globetrotter of Celestial Physics*, Astrophysics and Space Science Library 380, DOI 10.1007/978-1-4614-0697-6_11,

> I must tell you that already two English chemists have started to investigate my subject by repeating my experiments, but without affecting the square law, which they do not even understand, despite saying that they have confirmed it. I had been warned that they were to communicate something on the subject. I then demanded and was allowed to speak first. Their Communication, which they did not dare give themselves, was not brilliant. Everyone recognized the dishonesty of the action and the mediocrity of the work. These men did not bank on my presence at the meeting. [4]

He then comes to the famous phonograph:

> I have just received a visit from a representative and friend of Edison who brought me his portrait on his behalf, and also an autograph phonogram that he sent to me from America. This phonogram consists of a hollow cylinder of white wax, which was placed on the phonograph and on which the machine left traces under the effect of the author's voice. This is a friendly conversation that Edison has sent me across the Atlantic and which his friend suggests he will play to me on his behalf. The machine on which the cylinder will be placed is in transit, and we have arranged to meet in Paris about the 17th.

From Bath, Janssen sent the following message to Edison, which he recorded for him, and this was the first French voice to cross the Atlantic in such a new form: "The problem of artificially reproducing the human voice is one of the most amazing of those that Man could ever set himself. M. Edison's genius has provided us with the solution, and his name will be blessed by all those who are able to hear again the animated voices of those whom they would have lost" [5].

The Phonograph Under the Dome *[of the Académie Française]*

In reality, it was during the months of April and May of the following year, in 1889, that the Parisian sessions took place in increasing number. Actually, as Janssen stated in his presentation to the Académie des sciences on 23 April, Edison had wanted to add new improvements to the equipment that had been shown at Bath.

After the explanations given by Janssen, and the speech by the representative, Colonel Gouraud, a friend of Edison's of French origin, "phonographed" in front of the members of the August assembly, a whole series of recordings was made; as listed by *La Nature*:

> WORDS – The words of M. Janssen, past President of the Académie, addressed to M. Edison; the words of M. Berger to the same person; messages from the correspondents of several French newspapers, addressed to their editors.
>
> A few words in the following languages:
> French, English, Spanish, Italian, Dutch, Greek, Latin, Syriac, Turkish, Hebrew, and Arabic.
>
> MUSIC – The *Marseillaise*, played by the military band of the Queen's Guards; *Hail Columbia*, by the same, and *Marche du régiment*; a duet for piano and cornet, the music by Gounod; a cornet duet; *Ave Maria*, by Gounod, sung and accompanied by himself. [6]

According to Janssen's diary [7], it seems that the first session with the phonograph took place on 14 April, perhaps privately in the office of the Académie des

Fig. 11.1 *The duc d'Aumale, des Cloiseaux, Gounod, and Janssen testing Edison's phonograph at the Académie des beaux-arts, 27 April 1889.* Physique Populaire by Émile Desbeaux. © Author's collection

sciences, followed by another on the 18th, at Eiffel's home. On the 25th, it was at Janssen's Parisian apartment that the equipment was tested, to the great satisfaction of everyone present, although unfortunately Gounod was unable to participate. In his letter of apologies, he nevertheless indicated that he had already tested the equipment. But it was undoubtedly on 27 April, at the Académie des beaux-arts, that the supreme presentation took place. It was again Janssen who carried out the presentation, and the engraving that we have of the event shows that he appears very satisfied, with the hint of a smile, listening to his dear friend Gounod recording "Il pleut, il pleut bergère," under the attentive eyes and ears of the duc d'Aumale (a general and historian), and of the mineralogist Alfred des Cloizeaux (Fig. 11.1). The transcript of the session tells us even more:

- M. Janssen (testing the Phonograph):
- Demosthenes, Cicero, Bossuet, why are you dead? We could today hear your wonderful harangues from the very mouths of those who pronounced them!

Then a minor, very prickly point arises. Is the sentence correct, say some; is it accurate, the others say. Is it the mouths of the orators or the cylinder of wax that is repeating the harangues? M. Janssen resumes, tries to find the correction, and stops at the word 'de' or 'des'. The phonograph picks up the hesitation and the interruption. For a third time he repeats: "Demosthenes, Cicero, Bossuet ..." and a stifled laugh causes syllables to be missed. The phonograph faithfully reproduces the poorly stifled laugh. – It's perfect! it's astonishing! everyone cries.

Then the duc d'Aumale entrusts a few words to the Phonograph. And immediately, in a high, curt voice, the instrument speaks a sentence from the *Histoire des princes de Condé:* "Cavaliers de Gassion, sabre haut, pistolet au poing, se ruèrent sur l'ennemi ..." *["Gassion's riders, sabers high and pistols in their hands, fall upon the enemy ..."]* One could imagine hearing the commands to a regiment.

Finally, M. Ch. Gounod moves up and sings into the speaking trumpet: "Il pleut, il pleut bergère ..." and signs the song "Charles Gounod, member of the Académie des Beaux-Arts, of the Institut de France." And one of M. Gounod's colleagues conveys the general impression in these words: "That's a wax cylinder that, in a century, will be worth one hundred thousand francs." [8]

Oh woe! One hundred thousand times woe! The historical wax cylinder remains today as untraceable as the historical annular daguerreotype recording the transit of Venus!

On 8 May, a session was planned at Gounod's house with Janssen. Did it take place, and did those recordings survive?

The Phonograph at the Marmite

On 17 May, Janssen introduced Colonel Gouraud at a dinner of the Marmite, an extremely interesting republican society:

It was in the month of November 1873 that several friends met one Friday in a small restaurant at the Palais-Royal, and decided to meet again every month. Such was the origin of the Society, which took the name of Marmite *[cooking-pot]*, to clearly indicate that all the professions: literary, scientific, artistic, commercial, and others, gathered together at the same table in a truly republican spirit. Although becoming more numerous, the Society remained faithful to the spirit of friendliness and unity, as well as to the liberal views that inspired it. [9]

Under Janssen's presidency in 1889, the activities of the society were particularly varied. Indeed, dinners were given for the President of the Senate, to travelers returning from the Pamirs, to M. Guimet, the "founder of the Museum of Religions," and to the Vice-President of the Argentine Republic. Another was organized to glorify the Eiffel Tower [10]:

This nail, whose point has garnered so much glory,
Friends, do not doubt, it was made for us,
And I am going to tell you right away what it is:
To our President, it is an observatory![3]

Had Janssen not actually just carried out experiments by observing a powerful light installed at the top of the tower, using his spectroscope at the Observatory of Meudon, at a distance of 7,700 m, precisely because "the thickness of the atmosphere traversed represents the action of the terrestrial atmosphere on rays from the Sun near the zenith" [11]?

---

[3] Ce clou, qui sur sa pointe a massé tant de gloire,
Amis, n'en doutons pas, c'est pour nous qu'on l'a fait,
Et je m'en vais vous dire avant tout ce que c'est:
Pour notre Président c'est un observatoire!

The society had also visited the Toltec Museum at the Trocadéro, and then the Museum of the history of Religions, before it opened to the public; as well as the display of the Panorama of the Trans-Atlantic Fleet; and it was, of course, represented at the unveiling of the statues to two of its former Presidents: that of Paul Bert at Auxerre, and of that of Henri Bouley, a former President of the Académie des sciences, at the Veterinary School at Alfort. It was at its meetings, among others, that Janssen came across his cousin Henri Le Roux, his friends Pierre Savorgnan de Brazza; Jules Claretie, the administrator of the Comédie-Française; the lawyer Émile Durier (President in 1886) father-in-law of Henri Le Roux, and his brother Charles, President of the Club alpin français; Gustave Eiffel (President in 1890), and his son-in-law, the engineer Adolphe Salles; René Goblet, former prime minister; Ernest Javal, Director of the National Institute for the Deaf and Dumb (the brother of the great ophthalmologist and inventor Émile Javal, whom Janssen also met several times); and, naturally, the dearest of them all, the painter Jean-Jacques Henner.

It may be noted that in 1888, Janssen lent the hall in the Orangerie at Meudon for the Marmite's meeting and celebration of 29 June. The dinner took place under electric light, and the singer of the concert that followed was none other than Mlle Janssen …

The Phonograph at the Exhibition of 1889

Edison, who had traveled to Paris with his wife to visit the Great Universal Exhibition of 1889, was very pleased by Janssen's welcome and of the trouble that he had taken to inform the influential Academies of the young inventor's wonderful achievement. Witness to this is the letter of thanks that Janssen received in September:

> My Dear Mr Janssen,
>
> In taking leave of this beautiful city of Paris and in saying good bye to the host of friends which it has been my good fortune to find in France, my mind constantly reverts to the day which I spent with yourself and family in Meudon. For your great generosity and all your kind hospitality I am joined by Mrs Edison in offering you my warmest thanks, and I say Adieu to you in the hope that at some not distant day I may again have the honor of shaking hands with a man of whom all France is justly proud, and who had excited the admiration of scientists throughout the whole of the civilized world.
>
> I am, my dear Janssen,
>
> Yours very gratefully,
> Thomas A Edison
>
> Paris Sept 10 1889 [12]

References

1. Janssen, Jules, "Sur le phonographe de M. Edison", *CRAS*, **108** (23 April 1889), 833–5.
2. Letter from Jules to Henriette Janssen, 05 September 1888, BIF Ms 4133–296.
3. Letter from Jules to Henriette Janssen, 10 September 1888, BIF Ms 4133–297.

4. Letter from Jules to Antoinette Janssen, 11 September 1888, BIF Ms 4133–291.
5. Janssen, Jules, "Sur le phonographe de M. Edison", *CRAS*, **108** (23 April 1889), 833–5.
6. Hospitalier, É[douard], *La Nature*, No. 851, 04 May 1889, 363–6.
7. Jules Janssen's notebook, BIF Ms 4130–7.
8. Cited by Desbeaux, Émile, *Physique populaire*, Paris, Flammarion, 1891, p. 50.
9. *Annuaire de la Marmite, 1890,* Paris, Imprimerie Richard.
10. Sonnet by Lucien Pâté, 14 June 1889, in *Annuaire de la Marmite.*
11. Janssen, Jules, "Sur l'origine des raies telluriques de l'oxygène dans le spectre solaire", *CRAS*, **108**, (20 May 1889), 1035–7.
12. Letter from Thomas Edison to Jules Janssen, 10 September 1889, BIF Ms 4135–96.

Chapter 12
The Saga of the Observatory on the Summit of Mont Blanc

"L'entreprise est de celles qui passent l'ordinaire, tant par la conception que par l'exécution, et il n'est pas d'un médiocre esprit de pousser jusqu'à la dernière limite du possible."

[The undertaking is one of those that surpass the ordinary, both in conception and in execution, and not one that a mediocre brain could drive to the ultimate limits of the possible.]

Charles Durier [1] (Fig. 12.1)

Altitude Is Essential

Ever since the beginning of his studies of the solar spectrum, Janssen knew that he had to work at altitude to minimize the effects of the telluric lines, caused by the Earth's atmosphere. He went to the Himalaya in 1869, and he climbed Vesuvius again in 1904, but ever since he had his solar observatory at Meudon he had dreamed of a subsidiary station, within sledge's reach. The creation of Meudon had been the culmination of 18 years of discoveries achieved during multiple, difficult expeditions, which demonstrated to his country the imperative necessity for it to be founded. The observatory on Mont Blanc would be the culmination of the subsequent 18 years. Janssen's tenacity would not only allow him to convince sponsors, but also to defeat the mountain. The idea was revolutionary, and far in advance of its time. Others had already constructed observatories on rocky slopes: he would construct one on the summit (4,810 m), and knowing perfectly well that the inaccessible rock was more than 12-m deep beneath the ice!

The Unveiling of the Monument to Saussure

Janssen loved the mountains because he had understood a long time earlier that they are scientific allies. When he went to Chamonix on 28 August 1887, it was not to observe the solar spectrum in his research on oxygen, nor to photograph clouds,

F. Launay, *The Astronomer Jules Janssen: A Globetrotter of Celestial Physics*, Astrophysics and Space Science Library 380, DOI 10.1007/978-1-4614-0697-6_12,

Le Petit Journal

Le Petit Journal

Le Supplément illustré

SUPPLÉMENT ILLUSTRÉ

Huit pages : CINQ centimes

ABONNEMENTS

Douzième année DIMANCHE 4 AOUT 1901 Numéro 559

LES CHASSEURS ALPINS AU MONT BLANC

Fig. 12.1 *French alpine infantry on Mont Blanc. Janssen's observatory is on the summit.* Le Petit Journal, *4 August 1901.* © Author's collection as he would do in the following October at an altitude of 2,780 m on the Pic du Midi de Bigorre in the Pyrenees, but as a representative of the Académie des sciences at the unveiling of the monument erected in memory of Horace Bénédict de Saussure, the Swiss geologist and physicist. "With Bénédict de Saussure, who climbed Mont Blanc on 5 August 1787, accompanied by Balmat [...] science took possession of these high Alpine regions, where significant scientific observations and determinations were carried out with great precision" [2], was what Gaston Tissandier reported at the time in the journal *La Nature*. Janssen stayed at the Mont-Blanc Hotel, where he not only met his friends from the French Alpine Club again, but also the Minister for Public Instruction, Eugène Spuller, a loyal supporter of Gambetta, whom he had accompanied in a balloon to Tours. What is more, Spuller admitted to Janssen that he was "happy to be assisted by a friend" for the unveiling. Begged to speak at the banquet given that evening, Janssen described to his wife how he managed:

> I took the plunge. Happily, I knew what I wanted to say. I spoke against the far too great scientific specialization nowadays, and showed that it was desirable, without abandoning one special branch one should emulate de Saussure's example, and occasionally mount the heights to embrace the plains and the relationships of the sciences to one another; that the discoveries of spectral analysis had been produced by connections born of an overall view of Physics and Astronomy; that there was today a mass of points of contact, just as fertile, that could be established between the sciences; that the erection of a statue to the man who had embraced Physics, Mineralogy, Geology, and Botany seemed to come at the right time to view things as a whole. I said that because I found myself among Alpinists, I would call this intellectual alpinism, and I would inform my colleagues of this new field. I ended by toasting intellectual alpinism. [3]

There can be no doubt that at that moment, Janssen was already thinking very seriously about his future interdisciplinary observatory for expeditions on Mont Blanc, all the more so because, the previous year, a man called Joseph Vallot, a self-taught geophysicist, had made the first scientific ascent to the summit, and that in that very year, a month earlier, he had spent three consecutive days there, from 27 to 30 July, with a large number of meteorological, physiological, and photographic instruments.

The Ascent to the Grands-Mulets Refuge in 1888

Strengthened by this exploit, which none had thought feasible, Vallot decided to launch the construction of a wooden chalet, a laboratory sited on the rock at the foot of the Grande Bosse, 450 m below the highest point. During the course of the summer of 1888, he sent a letter to the Chamonix guides, without whose collaboration he would be unable to complete the project. It was precisely in October 1888 that Janssen, the new President of the Alpine Club (which should not surprise us), decided to undertake a scientific mission to the Grands-Mulets refuge, at an altitude of 3,050 m. That time of year was not chosen by chance. Indeed, Janssen always

Fig. 12.2 *Janssen in his "ladder-chair," from a photograph by E. Whymper.* La Nature, *19 November 1895.* © Author's collection

wanted to decrease as far as possible the amount of terrestrial oxygen and water vapor that appeared in the solar spectrum:

> This meant making observations during very cold weather. It was because of this that I was led to climb to a high station to reduce a considerable amount of the effects of atmospheric oxygen, and to choose cold weather to counteract the action of aqueous vapor, and to leave spectral effects in all their simplicity. [4]

The big problem for Janssen was that he could not walk for very long, because of his limp. So he needed to find a solution, and this was again an occasion for him to come up with a new invention: the "chaise-échelle" *["ladder-chair"]*. Here is the description of it that he gave:

> This device consists of a sort of ladder, some 3 to 3.5 metres long, the ends of which rest on the shoulders of four or six porters; the traveler is located between two rungs, in the centre, on a light seat suspended by straps, such that the uprights do not touch the armpits and the arms are free, outside them. At points where it is absolutely essential to walk, the traveler can place his feet on the ground without leaving his central position inside the ladder, and is then supported under his armpits by the device's stringers, which greatly lessens fatigue. If a crevasse is encountered, the ladder may be placed across it, making crossing easier. Finally, when circumstances require it, the traveler, wrapped up in thick blankets, can lay on the device and be carried at arm's length by his group of porters. [5] (Fig. 12.2)

The expedition left Chamonix on the morning of 12 October, "with mules for the traveler and baggage at far as Pierre-Pointue." But even on mule-back, the trip was not easy, and Janssen, who slipped from his saddle, had to have himself carried: it was Antoinette, who was of the party, who wrote to her mother when they arrived [6].

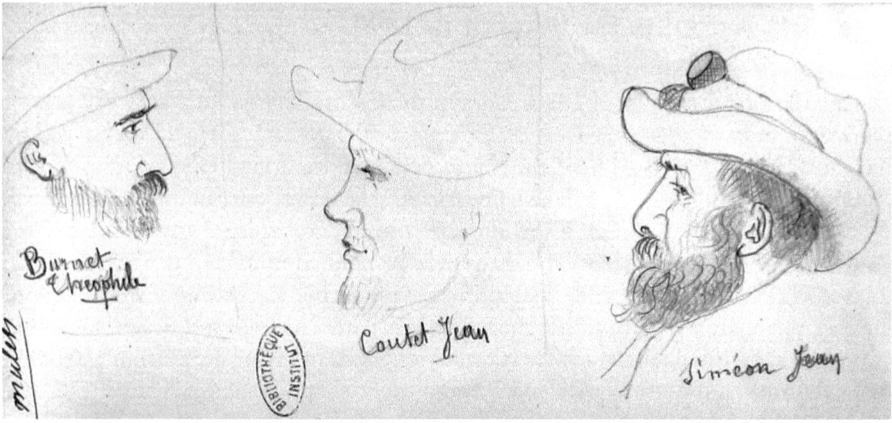

Fig. 12.3 *Three of the guides from the ascent to Grands-Mulets in 1888. A drawing by Jules Janssen.* © Library of the Institut de France

After a night spent at 2,000 m, the group started again on the 13th (Fig. 12.3). The "ladder-chair" was fully employed on the glaciers, but it still required 13 h for Janssen to reach the Grands-Mulets chalet as night fell, instead of the 4 or 5 h that were normally necessary. The next day, Janssen wrote to his wife: "Antoinette is well, although very tired, and she will probably go back down tomorrow. The good weather means that I shall remain. The beauties of the route are indescribable" [7]. Janssen made observations until 16 October under a wonderfully pure sky. When the Sun rose, the lines and bands of water vapor were completely absent from the solar spectrum, and when it crossed the meridian, some of the oxygen lines were absent. Of those that remained visible, Janssen deduced, by comparison with, and extrapolation from observations at Meudon that they were exclusively caused by the Earth's atmosphere. However, he remained cautious, and in this the future was to prove him right:

> Should we then conclude from this that oxygen does not enter into the composition of the globe of the Sun? When spectral analysis began, we might have been tempted to draw that conclusion; nowadays we have learned to be more cautious. The oxygen that might exist in the deep layers beneath the photosphere and the spots would not give any indications accessible to our current methods of spectral analysis. [...] This is a stage in the story of the relationship of oxygen with the Sun. It is an initial basis from which science can take its investigations further. [...] Yes, the mountain, and above all, certain specific mountains will play a major role in the astronomy that is to come. [8]

And Janssen continued, again evoking the matter of the search for extraterrestrial life, in which he was intensely interested:

> This Astronomy is extremely important and it will be very beautiful. With it, the science of the heavens will finally attack this great problem, which is as old as science itself, and which dates from the earliest reflections of Man about the celestial sphere and about the bodies that are its glory, namely: whether the bodies that we see are inhabited, whether life exists outside the Earth, and if beings similar to ourselves inhabit other worlds. [9]

The First Ascent to the Summit in 1890

During the first 4 months of 1890, Janssen worked in Algeria, in a very dry part of the world, in particular at Biskra "on the edge of the desert," but at a "site served by the railway" [10]. He had tried in vain to persuade his friend Henner to join him: "Let yourself be tempted. It will be a fine trip and the desert sun is a sovereign remedy for ills" [11], he had written a few days before his departure. Only Henriette and Antoinette joined him. Although when there he studied mirages, he wanted, above all, to observe the whole solar spectrum at the horizon. He was now working with another dispersion system rather than with a prism. This was a diffraction grating constructed by Rowland, which "is currently engaged in giving a spectrum extending from the red to the ultraviolet and of admirable accuracy and richness" [12], he explained. In addition, Janssen had available new photographic emulsions which allowed the yellow and red to be recorded. He stated with satisfaction that "the strength of the telluric lines is as low as possible" [13].

When Vallot organized the construction of his observatory-cum-refuge at the Bosses (at 4,365 m), which was finished 3 August 1890, Janssen, who was still aiming for the summit, took immediate advantage of it. This time, the chair carried by porters did not suffice but, because "there are very few difficulties that cannot be surmounted by determination and sufficiently deep study" [14], Janssen, who was then 66 years old, added a sledge constructed at Meudon Observatory:

> Its form generally resembles that of Lapp sledges, but I had a handrail, very solidly fixed, added along two-thirds of its length and towards the head [...]. I also had a long rope ladder made, with wooden rungs, which could be fixed to the sledge. [15]

The expedition, consisting of Janssen, his friend Charles Durier, the then Vice-President of the Alpine Club, and 22 guides or porters, moved off from Chamonix on Sunday, 17 August. The ladder-chair was used as far as the Grands-Mulets, where the night was spent. The next day, the sledge (Fig. 12.4) was used to reach the "cabin at the Bosses," which consisted of two rooms, overall measuring 5 m long by 3 wide and 3 high. Janssen set himself up in one of the rooms with Durier and Vallot, and tried to carry out some spectroscopy before a terrible storm, which reminded him of the one that occurred in Hong-Kong harbor in 1874, broke during the night.

> You cannot imagine the violence of the hurricane, mixed with snow and hail, which kept us confined for sixty hours, and which was none other than the cyclone, which remains famous in meteorological records (18–20 August 1890), which crossed part of France, devastated several valleys in the Jura, and in one minute, knocked down sixty thousand pines around Saint-Claude. [16]

On the 21st, Vallot left the cabin and went down again, extremely happy with the sentence inscribed the day before in the visitors' book of his laboratory-cum-refuge by Janssen: "M. Vallot has done science and alpinism a great service, the significance of which the future will reveal. The three-day stay that I made there with my friend Charles Durier has shown that one can make useful and regular observations at these great altitudes" [17]. Janssen remained. He repeated some "careful observations" with a Duboscq spectrograph and, on the 22nd, with superb weather, the climb could recommence. Charles Durier gives a particularly moving account of it:

Fig. 12.4 *Janssen in his sledge on the way up Mont-Blanc, 18 August 1890. From a photograph by M. Saladin.* La Nature, *27 September 1890.* © Author's Collection

Our team, reduced to twelve elite men, showed a wonderful spirit. I shall never forget the impression that I felt when, arriving at the top of the second Bosse, I turned round and saw below me, in plummeting perspective, the sledge, being pulled, pushed, and buffeted about on that acute crest, between two dreadful precipices. It was a supremely moving sight, and yet the vision contained, at the same time, something incomprehensible. The wind was lifting thousands of glittering flakes of snow, which whirled in the air like silver dust. To see men through this sparkling cloud, so bent over against the slope that one could not make out their faces, the legs, arms, and the ice axes that were waving around confusedly on each side of the ladder, as if along an enormous spine, and one would have said that it was some fantastic animal, a giant centipede, a monstrous millipede, fiercely climbing Mont-Blanc. [...] This second hump was, in reality, terrible: I had a great feeling of relief when [...] the team moved away from the precipices, and it was with intense joy that I shook M. Janssen's hand, whose face had remained impassive [...] At midday we attained the summit. M. Janssen took a Tricolor and waved it in view of Chamonix. Chamonix, from that height, was reduced to nothing; but what was strange was the flash from the viewing telescopes which, each time the objective was pointed exactly at us, shot out a spurt of light, and gleamed like carbuncles. [18]

Undoubtedly the eyes of the Janssen ladies were riveted to the eyepieces of some of those telescopes, and to whom Vallot had carried the news the day before. And when he reported the facts 3 years later, Janssen showed, as far as he was concerned, an Olympian calm:

> On Friday 22 August 1890, around midday, a sledge, driven, or rather hoisted, by twelve men with extraordinary vigor and resolution, reached the summit of Mont Blanc. It was the first time that the ascent had been made by such a new means, and that a traveler had reached the summit of this celebrated mountain, more-or-less without setting foot on the ground. The trip required heroic efforts: we had had to climb extremely steep slopes, cross or bypass wide crevasses, cross sheer ridges, bordered by precipices; and triumphed over all. [...] The traveler [...] was already casting his eyes on the scene taking place around him, and the view sent him into a sort of rapture or ecstasy. [...] The traveler did not allow himself to be completely absorbed in this unforgettable picture: he turned all his attention to the conditions that this site offered for astronomical and physical observations and, after this inspection and having taken some quick notes, he gave the signal for the retreat, and the troop went down again. The success of the trip showed that these glaciers, which offer to the scientist such new and significant subjects for study, and to the artist and the poet, pictures and scenes that are so grandiose and sublime, are henceforth accessible to those whose physical strength betrays their wishes. Let us also add that the method of traveling that overcomes the extraordinary fatigue of a climb on foot is the only form that enables one to fully enjoy the beauties of these high regions. [19]

As far as the spectroscope was concerned, when compared, the results from the three sites: Meudon, Grands-Mulets, and summit of Mont Blanc confirmed the absence of oxygen in the Sun's external envelopes, and that ensured "that our great central fire *[would have]* the greatest possible duration of the functions from which the life of the whole planetary system depends" [20]. Indeed, if there had been oxygen present, the formation of water by recombination with hydrogen would have resulted in a decrease in the heat radiation from the Sun ...

The Project, the Sponsors, and Gustave Eiffel

Not having had to expend any physical effort, Janssen was the sole person, "from what it appears, who retained the full integrity of his intellectual forces" [21]. As he explained, intellectual activity requires a great expenditure of energy, notably the presence of oxygen in the blood. It is thus necessary to husband one's physical forces if one wants to the capable of thinking at altitude. Janssen did not have to ponder long before deciding to construct an observatory at the summit. There, not only would the atmospheric pressure be reduced by about one half relative to what it would be at sea level (550 hPa instead of 1,013 hPa), but the horizon would obviously be much wider than on the slopes, and the air masses would, above all, be far more stable. The project took shape almost immediately, and on 22 September Janssen submitted it to the Académie des sciences. The following 27 July, he announced that his appeal to the "friends of Science" had been heard [22]. The sponsors were Raphaël Bischoffsheim, "to whom we owe the beautiful Nice Observatory," who provided finance to the tune

Fig. 12.5 *Gustave Eiffel.* Album Félix Potin. © Author's collection

of 140,000 F; Prince Roland Bonaparte, "who fosters geology with distinction," 100,000 F; and Baron Alphonse de Rothschild, of the Académie des beaux-arts, 20,000 F. Janssen gave 10,000 F, which, for him, represented a year of his salary as Director of Meudon Observatory, Mme Charles Heine, 1,000 F; Count Henri Greffulhe, 1,000 F; Count Aymar de La Baume Pluvinel, 1,000 F; and the state allocated an annual sum of 12,000 F, later reduced to 8,000 F [23]. The President of the Republic, Sadi Carnot, who received Janssen at Fontainebleau in August 1892 [24], even agreed to become an honorary member of the committee that was formed, while the former minister, Léon Say, was Honorary President. Léon Say was an old acquaintance, who, in 1887 invited Janssen (who immediately accepted), to join him in organizing the celebrations for the centenary of the Revolution:

> 100 years to you is nothing: and yet the star of the Revolution, despite being only 100 years old, risks entering a period of decline if we do not make efforts to restore to it the true brilliance that it can only find in the pursuit of liberal and moderate politics.
> I am sending you the notice of a display that we want to make at Versailles to enable us to take control of the direction of the centenary of 89, and remove it from the radicals.
> You will powerfully assist us by joining us. [25]

Finally, and this was by no means the least contribution, the engineer Gustave Eiffel helped, and assumed the costs of the preliminary studies, which amounted to 15,000 F [26]. Eiffel (Fig. 12.5), whom we have already encountered at the Marmite, was a friend of Janssen's, and the two families were very close. Photographs of Janssen, of his wife and Eiffel observing Mont Blanc from Chamonix with a telescope are to be found in the Eiffel archives, as well as a large framed print of Edward Whymper's photograph showing Janssen carried in his ladder-chair by porters (See Fig. 12.2), dedicated by Janssen to Mme Salles (née Claire Eiffel) in October 1899: "to the splendid daughter of the Great Engineer, as a sign of great regard and most affectionate respect" [27]. In 1893, Janssen lent his support to Eiffel, severely indicted in "the scandal" that followed the sensational bankruptcy of the Compagnie du canal du Panama *[Panama Canal Company]*, an accusation of which he was completely absolved. On 20 October 1893, Janssen wrote to Eiffel that he was "extremely affected by his distress" [28] and, presumably at the beginning of 1895 sent him the following carte de visite: "Justice has already started *[to be done]*, it will soon be complete, and my heart as a righteous man, a good Frenchman, and a friend, will be fully satisfied" [29]. He even went so far, as did his colleagues, to resign in April 1895 from the Council of the Order of the Légion d'honneur, in support. Eiffel was particularly grateful to the man that he called "master and friend":

> How much have I been touched by the lofty feelings shown by your letter, and how much have I been moved by the noble words in which they are expressed! It is truly comforting for me to hear them after all the horrible comments, arising from some through malice and envy, and from others through lack of thought and idiocy, but which nevertheless still leave cruel wounds. Against this clamor, all I have as palladium is the invaluable verdict of the Council of the Order of the Légion d'honneur, and as a consolation only the esteem of a certain number of individuals such as yourself, which causes me to regard the opinion of others with utter disdain. [30]

And in the last years of his life, Janssen regularly met the engineer in the month of June, on the occasion of the festival to the Sun, organized by his friend Flammarion… at the Eiffel Tower [31].

The Observatory's Inauguration in 1893

It was a Swiss engineer, Xaver Imfeld, who was given the task of drilling. Eiffel hoped to construct an entirely metallic building, topped by a dome, but he would not do so unless rock was found at a depth of less than 12 m. Starting at an altitude of 4,796 m, Imfeld had two horizontal galleries, 23-m long, and perpendicular to one another, excavated [32]. Alas! The workers only encountered hardened snow. Not the slightest rock! Never mind! Despite warnings, in particular that from Vallot, rivalry with whom was becoming more and more evident, and after preliminary studies which appeared to him to guarantee success, Janssen nevertheless decided to build his observatory on the ice. Initially, a small wooden kiosk left near the summit in September 1891 remained stable there for 15 months. In addition, Janssen had tests of the resistance of snow carried out in one of the courtyards of the outbuildings at Meudon: a column of lead, 35 cm in diameter and weighing 360 kg was placed on top of a pile of snow, one storey high, and to everyone's great surprise, it sank by just 7–8 mm. Janssen immediately calculated that a construction, with the base of 10 m by 5 m that he hoped to build, and weighing 187 t, would not even sink by a few centimeters at the summit of Mont Blanc [33]. As a precaution, a system of jacks

Fig. 12.6 *The Mont Blanc Observatory assembled at Meudon before transport.* La Science Illustrée, *1893.* © Paris Observatory Library

would also be provided, but as neither Eiffel nor Imfeld were any longer involved, it was a friend of Janssen's, the architect Émile Vaudremer, a member of the Académie des beaux-arts, who undertook, free, to draw up the plans for the building. It was to be a truncated pyramid, 7-m high, with double walls of deal, and which would first be assembled at Meudon [34] (Fig. 12.6). It was planned to have a living level, and an observation level, surmounted by a turret, sheltering a staircase.

The crated components were transported by the PLM railway company to Chamonix in 1892, and then, during the summers of 1892 and 1893 carried up on the backs of porters, with the help of the famous sledge and winches, specially designed by Janssen, to two intermediate shelters that had been constructed as depots, one at the Grands-Mulets and the other at the Rochers Rouges, at an altitude of 4,500 m. The porters had loads weighing up to 15 kg, that is twice their normal loads, and more than 15 t of material was carried up to the summit!

On 8 September 1893, the shell was finished, and Janssen, who was returning from a photographic meeting in Geneva, left for the summit. Henriette and Antoinette watched him through a telescope, although they complained that they did not have a suitable eyepiece. On the 9th, Henriette had a mass said at Chamonix, and advised her husband in writing not to leave it to others to announce his arrival at the summit and the report of the success of the opening [35] ... Actually, letters continued to pass between the town and the summit! Jules obeyed, and after having passed his first good night on the summit, "having made myself a bed in an alcove with blankets" [36], he sent the following dispatch to Prince Roland, on Tuesday, 12 September: "Observatory in place major work finished remainder still to be done. Winches used a lot for work and my successful ascent thanks to brave workers with favorable weather and to my dear collaborators so devoted to science, among whom you number greatly and I thank you again" [37]. What Janssen did not say then was that he was not happy with the work by the contractors, who had not sunk the construction as deeply into the snow as he would have wished. And what he did not yet know was that the weather was about to deteriorate, and that, having given preference to the instruments over food – unless the story as told by Fonvielle in the *Journal des Voyages* [38] (Fig. 12.7), who leads us to believe that certain porters willfully lost two crates to avoid having to carry them, is not a better reflection of reality – he had to go 2 days without food. That also meant that he had to wait until the Thursday to use his Rowland grating.

The new observatory soon garnered publicity, because when he returned to Meudon, Janssen received a visit from Ida Tarbell, who was to devote an article 13 pages long on the building, in the famous American publication *Mc Clure's Magazine*. In addition to details of the edifice and the science that Janssen wanted to carry out there, the journalist, obviously impressed by his character, gave the following description:

> To see M. Janssen, a little man of nearly seventy years of age, with a halt in his gait, going out in his library, where piles of books and pamphlets, astronomical contrivances and celestial photographs hide him half of the time from view, one would not at first admit the possibility of his living through so dangerous and exhausting an expedition as the ascent of Mt. Blanc. It is only when one remarks the freshness of his face, half-hid under the abundant snowy hair and beard, the brightness of his eye, and the youthfulness of his voice that one sees the real vigor and intrepidity of the man. [39]

The article appeared in February 1894 (Fig. 12.8), but friends such as Huggins had not waited for its appearance to congratulate Janssen:

> I am delighted to write to you again, above all to send you our most sincere congratulations on the beautiful realization of your great idea of climbing almost to the heavens themselves, by crowning Earth's summit with an observatory![1] [40]

[1] The original text was written in French: "Il m'est bien agréable de me mettre encore une fois en communication avec vous, et surtout lorsque c'est pour vous offrir nos plus sincères félicitations sur la belle réalisation de votre grande idée de monter presque au ciel même en couronnant le sommet de la terre par un observatoire!"

Journal des Voyages

ET DES AVENTURES DE TERRE ET DE MER

(SUR TERRE ET SUR MER; MONDE PITTORESQUE; TERRE ILLUSTRÉE réunis)

DIMANCHE 26 MAI 1901

Journal hebdomadaire. ABONNEMENTS : UN AN : PARIS, SEINE ET SEINE-&-OISE, 8 fr. — DÉPARTEMENTS, 10 fr. — UNION POSTALE, 12 fr. *rue Saint-Joseph, 12, Paris, 2e.*

N° 234 2e SÉRIE | DANS LES ALPES FRANÇAISES | M. JANSSEN AU MONT-BLANC | PAR WILFRID DE FONVIELLE | PRIX 15 c.

Au moindre faux pas, toute l'équipe était menacée de rouler dans l'abîme. (P. 410, col. 2.)

VOIR DANS CE NUMERO
Composition d'Edouard ZIER, continuant la série du

ÉMILE VITARD
N° 234. *(Deuxième série.)*

SANG GAULOIS
N° 1246 de la collection.

Fig. 12.7 *The least stumble, and the whole team was threatened with falling into the abyss.* Le Journal des Voyages, *26 May 1901*. © Author's collection

Fig. 12.8 *Janssen's ascent of Mont Blanc.* Mc Clure's Magazine, *February 1894.* © Author's collection

The Observatory in Action

While, in 1894, in his laboratory at Meudon, Janssen studied the spectrum of oxygen at high temperatures – because the Sun possibly included hot layers, an assumption of more than just minor interest – he had a "long-term meteorograph" installed at Mont Blanc. The instrument, whose driving clockwork could last for 8 months, was designed for measurements of pressure, temperature, humidity, and speed and direction of the wind. Janssen also had photographs taken to see if the observatory was moving. No movement was detected then.

In September 1895, at 71, Janssen made his third and last ascent to the summit (Fig. 12.9): he hoped to make new spectroscopic observations, see the broken-down meteorograph and, above all, to install the equatorial telescope, mounted with a polar siderostat, whose objective, with an diameter of 30 cm, and a focal length of 5.40 m, was a donation from the Henry brothers, as well as the 60 cm mirror of the siderostat (Fig. 12.10). In fact, the refractor, equipped for spectroscopy and photography [41], was not installed until 1896 (Fig. 12.11). While Bigourdan continued his measurements of gravity in 1897, Hansky, with Crova, restarted those on the solar-radiation

Fig. 12.9 *The observatory at the summit of Mont Blanc. Postcard.* © Author's collection

Fig. 12.10 *Assembly of the refractor for the Mont-Blanc Observatory at Meudon.* La Nature, *9 June 1906.* © Patrick Fuentes collection

Fig. 12.11 *The observatory at the summit of Mont Blanc, with the refractor visible at left. Postcard.* © Author's collection

constant. Janssen, who directed operations from Chamonix, where he arrived on a stretcher as a result of a severe fall that he suffered on the staircase of the large dome at Meudon, became the visionary again:

> It would require the sum total of six hundred thousand years of coal mining [...] to equal, in heat produced, the amount the Sun provides us with in a single year! This fact [...] shows us that when we come to turn to solar energy, we will have an immense source of power to put at the service of the needs of civilization. [...] When this great progress is realized, humanity will then be set on a truly rational and prudent path. Until now, we have lived on the reserves accumulated over the ages that have preceded us. Thus we have been imitating prodigal children who dissipate the patrimony of their ancestors, instead of living sensibly on their income. [42]

Until 1906, Janssen regularly reported on the work carried out at the summit. They did not solely concern astronomical physics, such as the study of solar oxygen by La Baume Pluvinel and Tikhov; photography of the solar corona and the observation of an occultation of Saturn by the Moon by Hansky; study of the surfaces of Jupiter and Saturn by Millochau; observation of the Perseids and Hertzian waves by Nordmann; or atmospheric physics like study of the green flash by Hansky; atmospheric electricity by Le Cadet and André; and of ozone by Doctor de Thierry. They also concerned fields as varied as the determination of the electrical insulation properties of glaciers and its application to telegraphy by Cauro and Lespieau; the

[2] An excess of red corpuscles in the blood, now known by the term "polycythemia."

bacteriology of snow by Doctor Binot; altitude hyperglobulia[2] in guinea pigs, white rats and rabbits by Guillemard and Moog; and even spectroscopic studies of blood carried out by Doctor Albert Hénocque, Eiffel's brother-in-law, whom Follin had put in contact with Janssen.

The End of the Venture

After a small settlement in 1895, leveling was necessary in 1904, and the jacks had to be used in 1906, while a separate refuge was constructed to serve as a shelter for travelers. While the Vallot observatory, which had slipped slightly in 1898, was perfectly stable, the summit ark began to sink more and more after Janssen's death in 1907, and it was decided to demolish it in 1909, as a safety measure. Only the turret was recovered and taken down to Chamonix, where it is still visible in the Alpine Museum. The venture, which was felt at the time to be rather a failure, appears to us today in a different light:

> With hindsight, when we examine the philosophy behind this extraordinary venture, Janssen appears, on the contrary, as a pioneer and we should not be astonished that he was not fully understood in his own time. He conceived his project as is currently done in modern astronomy (large, ground-based instruments, or even satellites), that is as a limited operation with a maximum working lifetime of some 10 years. In this view, the erection of an observatory on the ice, with the precautions that were taken, was justifiable. Stable for a decade, it enabled unique results to be obtained, thanks to the rarefied and pure nature of the atmosphere on Mont Blanc, results that could not have been obtained elsewhere. In this sense, the daring Mont-Blanc operation was a success. [43] (Fig. 12.12)

Fig. 12.12 *Stereoscopic view of the observatory at the summit of Mont Blanc. Photograph taken in 1901.* © Author's collection

References

1. Durier, Charles, *Le Mont-Blanc*, Paris, Fischbacher, 4th edition, 1897, p. 483.
2. Tissandier, Gaston, "Le monument de Bénédict de Saussure", *La Nature*, No. 746, 17 September 1887, p. 256.
3. Letter from Jules to Henriette Janssen, 29 August 1887, BIF Ms 4133–282.
4. Janssen, Jules, "Ascension scientifique au refuge des Grands-Mulets (Mont-Blanc)", *Le Tour du Monde*, 1st semester 1889, 337–44.
5. *Ibid.*
6. Letter from Antoinette to Henriette Janssen, 12 October 1888, BIF Ms 4133–292.
7. Letter from Jules to Henriette Janssen, 14 October 1888, BIF Ms 4133–299.
8. Janssen, Jules, "Ascension scientifique au refuge des Grands-Mulets (Mont-Blanc)", *Le Tour du Monde*, 1st semester 1889, p. 343.
9. *Ibid.*
10. Janssen, Jules, "Notes sur les travaux récents exécutés en Algérie", *CRAS*, **110** (27 May 1890), 1047–9.
11. Letter from Jules Janssen to Jean-Jacques Henner, 21 December 1889, Musée Henner archives, letter No. 37.
12. Janssen, Jules, "Notes sur les travaux récents exécutés en Algérie", *CRAS*, **110** (27 May 1890), 1047–9.
13. *Ibid.*
14. Janssen, Jules, "Compte rendu d'une ascension scientifique au Mont-Blanc", *CRAS*, **111** (22 September 1890), 431–47.
15. *Ibid.*
16. Durier, Charles, *Le Mont-Blanc*, p. 474.
17. Cited by Vivian, Robert, *L'épopée Vallot au Mont Blanc*, Paris, Denoël, 1986, p. 85.
18. Durier, Charles, *Le Mont-Blanc*, p. 476–7.
19. Janssen, Jules, "Quatre jours d'observation au sommet du Mont Blanc, Séance publique des 5 académies du 25 octobre 1893", in *Lectures Académiques, discours*, 113–29.
20. Janssen, Jules, "Compte rendu d'une ascension scientifique au Mont-Blanc", *CRAS*, **111** (22 September 1890), 431–47.
21. *Ibid.*
22. Janssen, Jules, "Note sur un projet d'observatoire au Mont Blanc", *CRAS*, **113** (27 July 1891), 179–81.
23. Vivian, Robert, *L'épopée Vallot au Mont Blanc*, p. 108.
24. Letter from Jules to Henriette Janssen, 22 August 1892, BIF Ms 4133–332.
25. Letter from Léon Say to Jules Janssen, 01 November 1887, BIF Ms 4137–31.
26. Durier, Charles, *Le Mont-Blanc*, p. 478.
27. Musée d'Orsay archives (ARO), Eiffel collection, ARO/1981.
28. Letter from Jules Janssen to Gustave Eiffel, 20 October 1893, ARO, Eiffel collection, ARO/1891.
29. Carte de visite from Jules Janssen to Gustave Eiffel, ca. 1895, ARO, Eiffel collection, ARO/1891.
30. Letter from Gustave Eiffel to Jules Janssen, 28 January 1896, BIF Ms 4135–105.
31. Faidit, Jean-Michel, "La fête du Soleil", *l'Astronomie*, **118** (2004), 370–3.
32. Janssen, Jules, "L'observatoire du Mont Blanc", *La Nature*, No. 963 03 November 1891, 371–8.
33. Janssen, Jules, "Note sur l'Observatoire du Mont Blanc", *CRAS*, **115** (28 November 1892), 914–9.
34. Tissandier, Gaston, "L'Observatoire du Mont Blanc", *La Nature*, No. 1062, 07 October 1893, 289–90.
35. Letter from Henriette to Jules Janssen, 09 September 1893, BIF Ms 4134–231.
36. Letter from Jules to Henriette Janssen, 12 September 1893, BIF Ms 4133–336.

37. Dispatch from Jules Janssen to Prince Roland Bonaparte, 12 September 1893, BIF Ms 4134–233.
38. Fonvielle, Wilfrid de, "M. Janssen au Mont Blanc", *Journal des Voyages*, No. 234, 26 May 1901, 410–1.
39. Tarbell, Ida M., "The Observatory on top of Mt Blanc", *Mc Clure's Magazine*, February 1894, 289–301.
40. Letter from William Huggins to Jules Janssen, 03 December 1893, BIF Ms 4135–328.
41. Millochau, Gaston, "La grande lunette de l'Observatoire du Mont-Blanc", *La Nature,* No. 1724, 09 June 1906, 23–6.
42. Janssen, Jules, "Étude de la constante solaire au sommet du Mont-Blanc",
43. Malherbe, Jean-Marie, "Les observatoires du mont Blanc en 1900", *l'Astronomie,* **101** (1987), 115–34.

Chapter 13
Literary Salons and Educational Problems

> *"Sortez, Messieurs [les poètes] de cet étroit théâtre qu'on appelle la Terre, emparez-vous des cieux et de l'Univers."*
>
> *[Leave, Gentlemen, this narrow theatre that we call the Earth, and seize the heavens and the Universe.]*
>
> Jules Janssen [1]

The Salon of Mme Adam

Given that Janssen was one of the "500 contemporary celebrities" listed in the first album issued by Félix Potin, we should not be surprised to find his frequenting the literary salon of the celebrated Mme Adam, known to literature as Juliette Lamber. Here is what Adolphe Badin has to say about the elegant and witty "grande dame" (Fig. 13.1 left):

> It was in 1871, immediately after the war, that Mme Adam opened her famous salon, in that sumptuous apartment on the boulevard Poissonnière [...]. Mme Adam not only had the rare personal qualities indispensable for this difficult role: an incomparable mind, perfect affability, exquisite charm and – her beauty; but she additionally possessed in the highest degree a rapid sense for political matters, and a complete and absolute devotion to progressive ideas. So the salon on the boulevard Poissonnière quickly became one of the most lively centers of the republican movement. Under the "Moral Order,"[1] it even became significant as a focus of resistance: it was there that, at the most critical times, troubled spirits came running to regain their equilibrium, that those who were annoyed came to seek calm, and the disheartened to find courage. The brilliant triumph of her ideas and friends, while adding considerably to Mme Adam's personal influence, did not stop significant changes to the composition of her salon. One of its principal attractions was not long, actually, in being removed from it as a result of the dispersion of the most eminent personalities of the movement, or of their absorption by official posts, which retained some in their cabinet posts,

[1] This refers to the so-called regime of Moral Order, proclaimed by Mac-Mahon in 1873, a conservative reaction following the Franco-Prussian war.

F. Launay, *The Astronomer Jules Janssen: A Globetrotter of Celestial Physics*, Astrophysics and Space Science Library 380, DOI 10.1007/978-1-4614-0697-6_13,

Fig. 13.1 *Juliette Adam*. Félix Potin Album (*left*). *Madame Henry Gréville*. Mariani Album (*right*). © Author's collection

> others in more or less distant embassies, and yet others in their presidential chairs. On the other hand, the artistic and literary world, drawn by Mme Adam's intellectual reputation and by her well-known love for the arts, gradually assumed a preponderant place in the house, without any party distinctions. Indeed, it would have been difficult to find in Paris, or in Europe, a similar environment, where all the guests had a celebrated name, or were, at least widely known; where on the same evening, one might come into contact with numerous politicians – the deputies, senators, and ministers – nevertheless remaining faithful to their friend of bad times; as well as with men of letters, poets, and dramatic authors such as Turgenev, Dumas, Daudet, Leconte de Lisle, de Banville, de Bornier, Paul Déroulède, Jean Aicard, Henry Gréville, Ulbach, Theuriet, and Claretie; painters such as Bonnat, Henner, Laurens, Carolus Duran, Jules Breton, Detaille, and Bastien-Lepage; sculptors such as Paul Dubois, Guillaume, Chapu, Falguière, Mercié, Millet; scientists such as Janssen, Robin, Berthelot, Lauth, Pouchet, Dr Labbé, Trélat; great industrialists or great financiers like Heine, Arlès-Dufour, Koechlin-Schwartz, Siegfried, Turgan, de Reinach, and many others whom we forget. [2]

We may recall that Mme Adam was not content herself with just holding a salon, and it was with "rare fearlessness" that she dared to create, in competition with the *Revue des deux mondes*, another "great political and literary journal, widely open to ideas of progress and liberty." The first issue of her *Nouvelle Revue* appeared in October 1879.

Janssen and His Friend Henner

We would have undoubtedly found at Juliette Adam's salon well-known friends of Janssen, such as the sculptors Paul Dubois and Eugène Guillaume and, above all, the painter Jean-Jacques Henner (Fig. 13.2). But among the people who frequented

Fig. 13.2 *Jean-Jacques Henner*. Félix Potin Album. © Author's collection

the salon, there were also Mme Henry Gréville (Fig. 13.1 right), a romantic novelist who had taken as her "nom de plume" her maiden name. Alice Marie Céleste Henry had, since 1857, been the wife of Émile Durand, known as Durand-Gréville. The latter was an art critic and a translator of Turgenev, and his interviews with Henner during the period 1878–1888 reveal precisely the painter's frequent encounters with Janssen, who would, in fact, be the last person to visit Henner before his death [3]. The meetings sometimes happened at Henner's studio on the place Pigalle:

> 8 November 1879. I arrived at Henner's studio around four o'clock, just as Janssen was leaving. [...] Henner has, at the moment, several portraits on the go. That of Janssen is almost finished. A three-quarter-face portrait, turned towards the viewer's right, beard and grey-white hair done in a wonderful tint, and with a very light touch. Background dark green. [4]

Other days, the meetings occurred at Meudon: "Sunday 16 November 1879. We went, Henner and I, to M. Janssen's at the Physical Astronomy Observatory at Meudon" [5]. Janssen went to meet the two men at the station. Later, during the visit, the discussions were brisk:

> Janssen having remarked that in the bas-relief by Rude *[on the Arc de triomphe]*, the woman who personifies war is showing too much movement, and is opening her mouth too wide. Henner: "but because she is giving a call to arms, she can't open it too much!" [...] Regarding the *Léda* attributed to Leonardo da Vinci. Janssen having asked if it is really by that master, Henner declares that it is a superb piece, and that it would not be surprising that the picture should be by Leonardo: "In any case, it has been executed by a very good artist." [6]

We learn various anecdotes about Janssen's regular visits to the Louvre with Henner: "Henner, having invited him to dinner, first took him to the Louvre ... and the closure of the museum arrived to inform the master and his companion that the dinner would have to be on another day" [7]. He was obviously a master who, quite exceptionally on his part, did not hesitate to put his studio at the disposal of his friend when he was absent from Paris [8], and who appreciated the fact that his pupil took note of his remarks:

> 18 November 1881. Returned from the country *[Durand-Gréville tells us]*, Henner describes to me in picturesque terms the unexpected arrival of the astronomer Janssen at Bernwiller (Henner's birthplace in Alsace), their excursions, a twenty-four-hour stay at the convent of Sainte-Odile, where Janssen made a pencil portrait of the mother superior, more than eighty years old. The first sketch was drawn in a rather cramped style; after a remark by Henner, Janssen started again, and did something that was pretty good. [9]

We dream of being able to listen to these gentlemen's readings. "From time to time I pick up my La Fontaine," Henner said to Durand-Gréville on 8 January 1885. "The other day, Janssen read *L'homme et la couleuvre* [*The Man and the Grass Snake*] to me; he reads very well, even though he is an astronomer ..." [10]

And again sometimes it was at Durand-Gréville's home that they met "just men," together with the actor Adolphe Dupuis.

Janssen and Education

The men knew each other, but the women also saw one another, because Janssen finished a letter that he wrote to Gréville by: "I hope that Madame Gréville is well. Please be kind enough to pass on to her my kind respects and the faithful memories of the ladies here" [11]. Mme Gréville did not only write novels. Following Jules Ferry's law of 28 March 1882, which "excluded from the compulsory program *[of instruction]* the education of any particular dogma" and, elsewhere, "gives pride of place to moral education and civics" [12], she published, that same year, *L'instruction morale et civique des jeunes filles*. This was one of the four textbooks that was placed on the *Index* by the Catholic Church [13], like that by Paul Bert [14], whom we have already met with Henner and Janssen at the Marmite. Paul Bert had more than one problem with supporters of the clergy, and he was not afraid to speak out about this to Janssen on the eve of his election to the Académie des sciences: "Things get more and more complicated, and more and more take on a political-clerical character. Above all, it is a question of preventing my going through" [15]. Janssen would certainly have not been surprised by the decision to put the work on the *Index*, and we can bet that he would not have approved.

Janssen was actually extremely concerned about education, and the manner in which it was given, particularly to girls. In a letter written to his wife from the Himalaya in 1868, and thus still under the Empire, not only that he was taking time to draw up a program for a school for young girls, but also particularly stressed *[his views]* on the subject:

> I approve of your being busy with Mlle Sanssine's School, but it is necessary that it should be laified *[sic for "secularized"]*. Sisters prevent giving children reasonable instruction. Prayers, sacred history, and catechism absorb everything [...] Mlle Sanssine will never be

truly head of her establishment and will have no credit for it, unless she sends the sisters away (giving them something for their institution and thanking them) and taking on lay *[female]* teachers. [16]

On 11 August 1901, Janssen gave a speech at the Meudon schools' prize-giving. He received a visit from the teacher of the town's secular school on 23 July 1902, and in July 1903, he was particularly concerned to donate, through his photographer Pasteur, two dozen copies of his large *Atlas de photographies solaires* [17] to the children of Meudon.

In his speech at the prize-giving at the Légion d'honneur's school[2] at Écouen in 1893, he said how much he approved of the education being given to young girls:

The demands of our current social life, the needs of all sorts that have been created by our lifestyle urgently require that the modern woman should play her part in professional work that has, until now, been exclusively reserved for men. As such, you owe particular gratitude [...] to the creators of this education that places you in a position to fulfill this new role, which is nowadays essential to ensure your independence, your dignity if you remain single, and to bring you, if you enter into a marriage, a precious addition to your joint resources. [18]

He particularly appreciated the recent instruction in commercial bookkeeping, the success of which had been astonishing. And how many times in his letters had he insisted to his wife that she should continue her learning and acquisition of culture, to read novels, and to perfect her English!

We shall see later that Janssen was closely linked to Frédéric Passy,[3] like him, President of the Association française pour l'avancement des sciences, "devoted to the cause of education in all its forms" [19] and with whom he shared his aspirations for peace: "Just as war has desired to create its triumphs, why should peace not create its own?" [20]

And furthermore, did not Janssen, in the evening of his life, eventually adhere to the Association for educational reform that had been founded by Baron de Coubertin, with whom he sat on the International Olympic Games Committee?

Janssen and the Académie Française

When Janssen frequented the literary salons, it was certainly not just as a spectator. He was cultured, and was a brilliant speaker. He was justifiably proud of his speeches, using his pen to great effect – when he was not improvising with gusto. He also made the most of it to promote as much as he could the language of Bossuet[4]

[2] The Légion d'honneur had (and still has) boarding schools for children and grandchildren of members of the order.

[3] Frédéric Passy was an economist, a deputy, and joint winner (with Henry Dunant) in 1901 of the first Nobel Peace Prize.

[4] A French bishop famous for his funeral orations, among which was one for Henrietta Maria, the Queen of England.

("what an orator: What a great artist! What nobility of thought and what greatness of expression" [21]), for which he had the greatest admiration.

> As far as I was concerned, I gave a long speech, deliberately, so that our dear Germans should properly understand the French language. I will tell you that it does me good to speak in public. With an orator, everything is in play: the body, the lungs, and the intelligence. Bringing to fruition one's thought and the way in which it should be expressed is highly favorable to the development of our faculties. I find myself fully refreshed and renewed after a speech that I have properly given. [22]

We may recall the moral success gained at Washington, thanks to his eloquence.

In 1903, he published through Hachette *Lectures académiques, discours*, pieces chosen from his writings that he considered as the most significant, and which he sent to all his friends. They were delighted as, for example, Camille Saint-Saëns:

> Your beautiful book is a joy to me; [...] the details of your escapades on mont Blanc have given me vertigo; I don't see myself undergoing all that. And yet, I would have so much liked to see such marvels and to listen to the "fearful silence" of which you speak! it is too late! But you describe it so well that I come to imagine that I have seen it. [23]

Among the texts in the anthology, there is one that sets itself somewhat apart from the others: it is an essay entitled "Science and poetry," which Janssen read at the annual public session of the five Académies on 25 October 1902. He went back to primitive Man, who "in the presence of images of Nature, have described for us their impressions, and those impressions gave rise to Poetry." Then he saw how science was born "in Assyria and Egypt, if one considers only the methodical and sustained observation of astronomical phenomena to be able to predict their return; but only in Greece if we require a true scientific spirit, intending to study phenomena so as to deduce their causes." He therefore considered that "We need to go back to Greece if we wish to find the indisputable feeling for this union of Science and Poetry. The Greek genius, so supple, so penetrating, and so extensive, was destined to sense it and feel it profoundly. But after Greece, what a long interregnum or, rather, what a night for pure science and the speculations that arise from it." After having rendered energetic homage to former greats, from Lucretius, Cicero, Pythagoras, and Plato to Laplace, Buffon, Lavoisier, and Dumas, via Kepler, Galileo, Descartes, Newton, and Leibnitz, he finished with a lyrical flight of fancy that should really be read aloud, declaimed in the theatrical and vibrant tradition of major speeches that disappeared with Malraux[5]:

> Yes, it is Science that contains the Poetry of the future. That is why I would invite our young poets to seize this great domain. Leave, Gentlemen, this narrow theatre that we call the Earth, and seize the heavens and the Universe. If contemplation of our terrestrial world has already given birth to so many beautiful and sublime poetic and philosophical insights, what will it be when the whole Universe, with its sights, its laws, and its harmonies is offered for inspiration! It should be proclaimed out loud: Poetry would be transformed,

[5] As minister of culture, André Malraux delivered such a speech in 1964, when the ashes of the great French Resistance worker Jean Moulin were transferred to the Pantheon in Paris.

Fig. 13.3 *Armand Sully-Prudhomme*. Félix Potin Album. © Author's collection

> I should say transfigured. The human soul will then be raised, through you, to heights that will give it a feeling for a completely new moral dignity, which even responds to the great mystery of its destiny. [24]

Janssen's speeches always enthralled his listeners. Following his brilliant performance at Vienna in 1883, Forster from Berlin "who is not gentle with us French" said to him "What beautiful language!"; "his [*Forster's*] coldness was overcome," and did the Swede Gylden not make a long detour to say to Janssen "You spoke not as a member of the Académie des sciences, but like a member of the Académie française" [25]? And so the idea gained ground, and it is without surprise that we see Janssen undertaking the traditional "visits" with an aim to his eventual election. On 29 March 1903, he met José Maria de Heredia who was "very cordial," and then Jules Lemaître on 17 April [26]. Jules Claretie and François Coppée were too busy to meet him. Pierre Loti was also absent from Paris, but we may believe in the sincerity of the letter that he sent him:

> How much do I regret that my distance from Paris deprives me of receiving the traditional visits, and thus to have the honor of knowing you. You cannot be unaware that the academic votes always depend on very complicated combinations, which I cannot predict from this distance, above all from always being absent. Allow me, however, to tell you that I should be very flattered to see you among us. [27]

With Sully-Prudhomme (Fig. 13.3), things were more definite: the two men knew one another well and liked one another. Sully-Prudhomme, who had studied for the Polytechnique, but had to give it up because of eye problems, was a friend of Henri Schneider, whom he knew at Le Creusot, when he worked in the foundry administration in 1858. Janssen never failed to send him a copy of the *Annuaire du*

Bureau des longitudes in which he published notes, and the academician, the first winner of the Nobel Prize for Literature, regretted in his old age that his state of health prevented him from going to visit Janssen at Meudon from his house in the countryside outside Paris at Châtenay-Malabry where he then lived. In 7 April 1903, he therefore wrote to Janssen:

> As much as I am happy to learn that you have not renounced such a legitimate wish as to be one of us in the Académie française, I am embarrassed and perplexed in my predictions concerning the forthcoming elections, so great is the number of candidates (of whom several have already received pledges of success in the near future) that is announced today. It would be necessary to sound out the intentions of my colleagues, and I am not favorably placed to do this, given that I do not belong to any coterie and that I have never been approached by any group of voters in favor of this or that candidate. It follows that I do not know about any groupings. Is the Académie ready to add a third member of the Académie des sciences?[6] I cannot claim that it is so; I do not dare to share all the confidence shown by your advisers in this respect. I would have scruples, however, in that my uncertainty might have caused a delay in your candidature. What I do not doubt is that your competence in literature (which dates, I believe, from your youth) would make your participation particularly precious in our functions as a literary jury. [28]

We can see that Janssen's chances of success were very small, even if the future psychoanalyst Marie Bonaparte, the daughter of his friend Prince Roland, sent her greeting from Saint-Raphaël where she was staying: "At present there is talk of the elections to the Académie, I know perfectly well whom I would like to see there: the one who merits it so greatly ..." [29].

The end of this story is that Janssen, who had actually stood for election on 18 April 1903 to one of the seats left vacant by the deaths of MM. Gaston Paris and Legouvé [30], decided on the following 11 June to withdraw his candidacy [31]. A week later, the new immortals were to be the historian Frédéric Masson and the novelist René Bazin.

References

1. Janssen, Jules, "Science et poésie", in *Lectures académiques, discours*, Paris, Hachette, 1903., p. 61.
2. Badin, Adolphe, "Madame Edmond Adam, Juliette Lamber", Note extracted from *Étrennes aux Dames pour l'an 1882.*
3. Press cutting, 1905, BIF Ms 4136–318bis.
4. Durand-Gréville, Émile, *Entretiens de J.-J. Henner*, Paris, Lemerre, 1925, pp. 59–60.
5. *Ibid.* p. 67.
6. *Ibid.* pp. 68–69.
7. *Ibid.* p. 117.
8. Letter from Jean-Jacques Henner to Jules Janssen, 1881, BIF Ms 4136–7.
9. Durand-Gréville, Émile, *Entretiens de J.-J. Henner*, p. 133.
10. *Ibid.* p.232.

[6] The other two members in question were Poincaré and Berthelot.

11. Letter from Jules Janssen to Émile Durand-Gréville, 13 September 1901, Françoise Launay collection.
12. "Circulaire adressée par M. le Ministre de l'Instruction publique aux instituteurs, concernant l'enseignement moral et civique", 17 November 1883.
13. Decree of the Sacred Congregation of the Index of the 15 December 1882, published in the *Semaine religieuse* by Pamiers, 25 January 1883, AN F^{19} 6088.
14. Bert, Paul, *L'instruction civique à l'école,* Paris, Picard-Bernheim et Cie, 1882.
15. Letter from Paul Bert to Jules Janssen, 31 March 1882, BIF Ms 4134–293.
16. Letter from Jules to Henriette Janssen, BIF Ms 4133–111.
17. Jules Janssen's notebooks, BIF Ms 4132–6 and 4132–7.
18. Janssen, Jules, "Discours prononcé le 27 juillet 1893", in *Lectures académiques, discours,* 269–75, p. 272.
19. *Dictionnaire biographique des notabilités de Seine-et-Oise*, Paris, Henri Jouve, 1893.
20. Janssen, Jules, "Discours prononcé au 13e dîner de la Conférence «Scientia»", in honour of M. G. Eiffel, Saturday 13 April 1889, in *Œuvres scientifiques*, Vol. II, 129–36.
21. Letter from Jules to Henriette Janssen, 18 September 1883, BIF Ms 4133–272.
22. *Ibid.*
23. Letter from Camille Saint-Saëns to Jules Janssen, 04 May 1904, BIF Ms 4137–26.
24. Janssen, Jules, "Science et poésie", p. 61.
25. Letter from Jules to Henriette Janssen, 18 September 1883, BIF Ms 4133–272.
26. Jules Janssen's notebook, BIF Ms 4132–7.
27. Letter from Pierre Loti to Jules Janssen, nd. *[1903]*, BIF Ms 4136–385.
28. Letter from Armand Sully-Prudhomme to Jules Janssen, 07 April 1903, BIF Ms 4137–51.
29. Letter from Marie Bonaparte to Jules Janssen, 28 April 1903, BIF Ms 4134–316.
30. Letter from Jules Janssen to the permanent secretary of the Académie française, 18 April 1903. Académie française archives.
31. Letter from Jules Janssen to the permanent secretary of the Académie française, 11 June 1903. Académie française archives.

Chapter 14
Janssen and Communication

"C'est donc à la science de faire connaître à la société nouvelle sa grandeur et son utilité."

[It is therefore for science to make its grandeur and utility known to the new society.]

Jules Janssen [1]

The Société Philomathique and the Académie Des Sciences

From 1863, Janssen was a member of the Société philomathique, an interdisciplinary scientific and philosophical society, whose motto was "study and friendship," and which traditionally played the role of an "antechamber" to the Académie des sciences. It was there that he made his debut, under excellent conditions, which was what he emotionally evoked in 1888, at the time of the society's centenary:

> Yes, Gentlemen, I have retained a grateful memory towards a society that allowed me to be in touch, on the basis of an indulgent equality, with men such as Claude Bernard, Foucault, Bour,[1] and so many other eminent minds. How delightful and fruitful were those evenings that I then passed in the old place on the rue de Nesles, when, in just a few moments, I saw the highest and most diverse questions elucidated, in a delightful conversational manner, with that freedom, even that casualness, which gives the company of genius a double charm: that of seeing it, so to speak, in informal attire and to be present at the very operations that lead to the discovery of the truth, and also to be benevolently allowed to take part! What a thrill for a young scientist! [2]

And Janssen went on to praise the confidentiality of the society, which allowed a freedom "which one no longer has in the Académie." However, it is perfectly clear that he always had his sights set on the Académie des sciences and, as soon as elected, he rarely missed one of the weekly meetings, where he gave more than 110 papers, not to speak of the other tasks that he took on there.

---

[1]A graduate of both the Polytechnique and the École des Mines in Paris, Edmond Bour made many significant contributions to analysis, algebra, geometry and applied mechanics despite his early death.

F. Launay, *The Astronomer Jules Janssen: A Globetrotter of Celestial Physics*, Astrophysics and Space Science Library 380, DOI 10.1007/978-1-4614-0697-6_14,

The Meetings of the British Association for the Advancement of Sciences and Friends on the Other Side of the Channel

Janssen attached a great significance to communications, and he understood very quickly that it was necessary to travel to become known, and that, in particular, he needed to go abroad. His presence was always more effective in speech, rather than in writing, especially when he spoke in French. It was most probably in 1866, in Nottingham, that he first attended a meeting of the BAAS. It was then the 36th meeting of the British society, and until the 1880s, Janssen regularly crossed the Channel at the end of August or the beginning of September, unless he was on some distant expedition. During the last 20 years of his life, he preferred to pass the summers at Chamonix, in sight of Mont Blanc. It may be noted that he went there by train, with Henriette and Antoinette, and his collaborators who would be working at the summit, everyone benefiting from the half-price travel permit that Janssen requested every year from the Director of the railway company [3] ... Nevertheless, he visited Bath in 1888, but that was undoubtedly because of the phonograph.

These reunions were naturally the occasion not only for personal meetings at the venue itself, but also for visits, and Janssen particularly praised this change:

> Gentlemen, at present new routines are being established in the world of science, which will have the happiest consequences, not only for the progress of science itself and for the relationships between scientists, but even for international relations. I am speaking of those friendly visits that occur on the occasion of Conferences or major scientific or literary meetings. [4]

As far as visits are concerned, Janssen made an incredible number after 1866. It was actually at Nottingham that he met Huggins for the first time, and that the respect, which rapidly turned to friendship, was established. It was on that occasion that he was able to visit the solar observatory at Kew. The meeting at Exeter, in 1869, marked the start of the friendly relations that were established between the Janssen and Lockyer families. In 1871, before going to Edinburgh, he told Henriette that he had met his friend Williamson, Wheatstone and his daughters, and De La Rue. In 1873, he made the journey from London to Bradford with Huggins. The two friends thus had the chance to speak about the "certificate for the grade of Commandeur"[2] for Huggins, which had not emerged from the French administrative offices, because they had not been able to find Huggins' postal address. Janssen would have soon straightened that out, and at the same time, have Huggins and Lockyer elected as corresponding members of the Académie des sciences: "I intend" he wrote to Lockyer "to support your candidature for the Académie as correspondent, as well as that of Huggins, because I believe that you represent the two great aspects of celestial spectral analysis, and that the Académie cannot do better than to forge links with

[2] We consider this worthy of mention, although we have not been able to establish which order it refers to.

two such eminent men" [5]. Janssen informed Lockyer of his election by telegram on 29 December 1873, and Lockyer replied immediately:

> Je vous remercie de vos soins et de l'honneur que la France m'a faite grâce à vous. Je tacherai d'en être digne mais ça sera difficile parce qu'elle est si grande.
>
> Je suis très sérieux mais en Français il est difficile de dire ce que je veux dire. [6]
>
> *[I wish to thank you for your efforts and for the honor that France has given me, thanks to you. I shall attempt to be worthy of it, but that will be difficult, because it is so great.*
>
> *I am very serious, but it is difficult to say in French what I would like to say.]*

A few weeks earlier, Lockyer had sent Janssen a copy of his *Contributions to Solar Physics* [7]. He dedicated his work, both to Balfour Stewart, the future Director of the Kew Observatory, and to Janssen. "Encouraged by one friend I undertook the work which has brought me the other."[3]

At Bradford, Janssen attended a lecture by Williamson on "Transcendent Chemistry,"[4] and adds that he congratulates himself for having made the trip:

> These meetings are so profitable to me from every point of view that it is absolutely necessary for me to attend and that I should remain until the end. Especially in view of my expeditions, here I get all sorts of information about China, Burma, and the Cape. – About work in astronomy and geography, and from which I derive the greatest benefit. [8]

He also took advantage of the occasion to live in the English fashion, which was helped by the fact that people attending the meeting were not lodged in a hotel, but in the homes of "hosts," and he also attended a religious service, celebrated by the Archbishop of York:

> The presence at Bradford of the most famous scientists in England and from abroad, gave the Archbishop, who is considered to be a distinguished thinker and somewhat energetic, and who, moreover, as a member of parliament, is involved with current *[political]* issues, a unique opportunity to attack bad science, that is to say that which does not consider itself subject to the Church. So he therefore attacked, not science, but some scientists, led astray by their pride, who do not recognize the limits that God has set upon Man's intelligence. As the orator was the master of the choice of the grounds *[for argument]*, of the questions and the definitions, and no one could make any rejoinder, his arguments had a rather triumphant form – too triumphant even – because one could sense from the demeanor of many of the scientists in the congregation, that the preacher was castigating men of straw while the real adversaries were elsewhere, judging the restrictive topic to be ridiculous, smiling imperceptibly at the Archbishop's scientific pretensions. [9]

However, he did not forget science, and was pleased to have had the chance to see the largest refractor in the world:

> I have been here since yesterday, with M. Newall at Newcastle. As I have said to you, M. Newall is a rich Englishman, who has had built the largest Refractor in the world. This giant of Celestial Optics has an aperture of 25 English inches *[63 cm]* and a focal length of

[3] Cited by Lockyer's second wife and one of his daughters in their book that appeared in 1928: *Life and Work of Sir Norman Lockyer.*

[4] Transcendent chemistry deals with the substances that exist in the wide expanse of the universe, and with the laws that they obey.

> 30 feet *[9.15 m]*. The dome that encloses it is as large as a house. It is about 15 metres in diameter, almost the width of the front of our house, and everything is in proportion. The observer is carried on a movable platform, raised by a mechanism.
>
> Unfortunately, the weather at Newcastle is so frequently bad that this magnificent instrument is hardly used. I was lucky enough to have good weather to try it. All day yesterday we studied the Sun, either the spots, or the prominences and, that evening, as the night was very fine, we let ourselves be led on until three o'clock in the morning. I am going to stay a few days to see properly what one can do with such large instruments. [10]

Janssen undoubtedly met Alexander Herschel at Newcastle, where the latter was a lecturer, because Alexander's sister, Isabella, several months later sent to Janssen, on his behalf, the autographs of her father, Sir John, and her grandfather, Sir William Herschel. As one may imagine, these rare "samples" were particularly valued, and they illustrate the extent of Janssen's admiration for those two great astronomers.

On returning to London, the day of the famous visit to Airy about the revolver, Janssen spent the day with De La Rue, who showed him "his workshop and a very fine house at 73 Portland Place; an almost princely residence, where there were quite a number of rooms and luxurious furniture of unheard-of comfort – He showed me my bedroom, as he called it, for when I came to London. Yesterday I dined with Hofmann, the celebrated chemist" [11].

In 1875, at the time of the meeting in Bristol, he went to see Sir William Thomson, who talked to him about his work on the tides. He made an excursion to Stonehenge, and also visited the observatory at Oxford, after having spent a day with the physicist Sir William Grove, who wanted to see him. The circle of relationships was expanding, although that particular scientist was getting older and no longer carried out much science. At any rate, the next year, Janssen was nominated as Vice-President for the meeting at Glasgow.

In Dublin, in 1878, Janssen had the "good luck" [12], to meet Henri Martin, the historian, with whom he struck up a friendship immediately. They dined together the next day at the home of the President of the Royal Society, with Williamson and William Thomson, with whom he got on extremely well. The next day he visited the workshops of Howard Grubb, the great instrument manufacturer. The following day, he was given an honorary doctorate by Dublin University, which meant he had to dress in a scarlet gown. But the great event of the day was, above all, the visit to Parsonstown to see the largest telescope in the world, that of Lord Rosse (William Parsons), and Janssen was the guest, for one night, of his son, Lawrence Parsons, the fourth Earl of Rosse [13]. That was on 20 August, and the visitors' book of the famous place does indeed preserve the trace of Janssen's visit as observatory Director.

The following year, Janssen stayed with the leading master cutler and steel manufacturer in Sheffield. There, as a special mark of attention, the servant who brought him his breakfast spoke to him in French! At the works, he saw 10 t of Bessemer steel made in half-an-hour. However, his absence did not prevent him from directing Meudon from a distance, and Arents and his colleagues in charge of the upkeep and the administration of the observatory were asked to draw up a report every day [14].

With Huggins, the harmonious friendship had become remarkable. Did Janssen never dare to send the letter, the draft of which he preserved, or did he rather like the idea of retaining the draft of this unexpected letter, which was written in 1879, and which shows how much he admired him *[Huggins]*?

> Dear Huggins,
>
> Modesty forces me to confess to you that I am not the great artist represented with so much humor on the card that you sent. I have never seen so many beautiful things on our satellite. I have been content with a few, far more modest, observations, and to which, however, I attach a far greater price. And yet those observations are no more than a pleasant dream that I have had recently. But as this dream is related to the artist, I want to tell you about it, but beg you to keep my secret for fear of being mocked. So, having gone to bed after having thought about the constitution of the Universe, and decided passionately to make progress in such beautiful and such lofty fields of knowledge, I fell asleep and soon dreamed that I was transported to a neighboring planet. I started to travel around on it. Because no one spoke the same language as me, I did the best I could with signs, and finally after some difficulty I could more or less make myself understood. I intimated that I wanted to know if this land had philosophers and astronomers. I was told that the land did have several very famous ones, but I was informed that above all there was one superior philosopher, whom I was assured was very clever in several fields of knowledge, notably astronomy, physics, botany, archaeology, etc., and I wanted to see so universal a man, and had myself shown to him. He was happy to see me, and was kind enough to respond to my inquisitive questions. I pursued this as far as he allowed, and I became more and more impressed by his great faculties. He told me that, by means of a delicate separation of the elements of light, he had come (together with another scientist, one of his friends) to recognize that the material that forms most of the suns visible in the heavens was of the same nature as that in our Sun, and even in our Earth. Such a result staggered me, and I felt myself really small before such a man. He then informed me that he had been able to apply his analysis to comets and even to those clusters of light that he called nebulae, and in which he had, before all the philosophers on that planet, recognized that subtle air that comes from swamps and which gleams as will-o-the-wisp. All these results staggered me. I grew more and more attached to this superior being, and he was kind enough to continue to talk to me about his work. One day he assured me that he had succeeded in measuring the movement of celestial bodies, solely by subjecting the light that they send towards us to his incredible analyses. Another day it was the heat of the stars that he measured, as if it were from a furnace; yet again he dissected and separated the light from the stars, and printed it in an indelible manner. And when I add that the company of this great philosopher was as agreeable as his science was profound and marvelous, and that he knew how to joke and even had fun doing so.[5] [15]

Although Huggins had a modest fortune, which Janssen did not have, the careers of the two men were much alike: Huggins did not go to study at Cambridge as he would have wished, because he had to take over his father's business, and he went from being a very well-informed amateur astronomer to that of an astronomer, recognized by the like of the Royal Society. As with Janssen, the turning point for spectroscopy came with the work of Kirchoff and Bunsen. Like him, he soon constructed a spectroscope to study the spectra of stars. He also did a lot of spectroscopy in the laboratory, which is why he collaborated with William Miller, a professor of chemistry at Kings College. His work continued with studies of the spectra of nebulae, and he soon tackled those of comets.

---

[5] The draft stops here.

On returning from Bath, Janssen took advantage of his passing through London to take a look at the British Museum, the National Gallery, and to go and see again his close friends Lockyer, Huggins, and De La Rue. In addition, he visited Christie, with whom he got on well:

> I have been to see Christie, the Director of Greenwich Observatory. Madame Christie was very kind, and kept me to dinner. Christie showed me the observatory's instruments, and what he is doing in mapping the sky. Madame Christie showed me her garden, where she has flowers, fruit and vegetables. I was made to eat grapes from the greenhouse, *[where the bunches]* had enormous, very sweet grapes. We spoke about the observatory's organization. Christie complained of the government, which is nit-picking and parsimonious. He is happy with his Council, which is made up of men of science. It is the same thing more-or-less everywhere: politics takes precedence over everything. [16]

The French Association for the Advancement of Science

The BAAS, however, had gained a following, and an equivalent French association *[L'Association française pour l'avancement des sciences (AFAS)]* saw the light of day in 1872, in the spirit of the country's moral renewal [17], which was not so distant from that of the Marmite. Janssen attended the first meeting, which was held at Bordeaux, and where he met his colleague Respighi, and liked to mention that beginning at the British meetings. At Bristol in 1873 "I spoke in French *[he said]* and finished by a few words in English on the union, the health, and the brotherhood of the old British association and its young sister: the French association. – Prolonged cheers" [18]. He reiterated this the following year in Glasgow:

> I gave a speech in French on the scientific borrowings between France and England. – Universities copied by England – and the Association by France, and I finished with the toast *[in English]:* I propose a toast at union and confraternity of British Association with its young sister the French Association for the advancement of Science. – Hurrah, hurrah, hurrah!! [19]

In 1881, at the meeting in Algiers, he again met Henri Martin and Frédéric Passy, with whom he made an excursion to the Great Kabylia. He was then Vice-President of the Association, which meant that he was to preside over the meeting at La Rochelle in 1882, and he was pleased by the election of Passy as Vice-President.

Janssen's position in the AFAS was slightly different from the one he had at the BAAS. When on the other side of the Channel, he was spreading the most recent discoveries that had been the subject of papers at the Académie des sciences to foreign colleagues. He wanted to become known and, above all, recognized, and he built up his reputation as a researcher. At the AFAS, his position was one of an established researcher. His reputation was well established, and he played the part of a broadcaster of knowledge:

> Today, nothing significant can exist or develop without the support of public opinion. It is therefore for science to make its grandeur and its utility known to the new society. [...]

> Now, to conquer opinion, it is necessary for science to be constantly in touch with the public, by every accessible means, and it is from this point of view that I want to consider public lessons, lectures and scientific soirées; I see this as far more than intellectual development of a high order, I see it as propaganda, which seems to me to be urgent today. [20]

His natural eloquence ensured the success of the public lectures that he gave at Lyon, Paris, and Toulouse, as part of the Association.

There, he spoke, but he also showed other material. In 1877, the Hugginses came to the meeting in Le Havre, where Janssen produced his first, very-high-quality photographs of the Sun:

> Today, I gave my paper on the solar photographs and the spectra. It was M. Huggins, above all, who appreciated this, and he was astonished at the perfection of the photographs, a perfection that he had not expected, and which he found quite extraordinary. M. Huggins has actually come here with his wife. I had lunch at their hotel, and I subsequently invited them to mine, and we have visited various premises together. They have come to France on a pleasure trip or circular tour at a reduced price, which allows them to see five or six towns in the north of France during their excursion. [21]

In his letter, Janssen emphasizes the collaboration between the Huggins' couple [22], and which is still cited nowadays as an example:

> Madame Huggins is completely involved in the work of her husband, whom she helps as an <u>assistant</u> and which she sticks to like a leech. It is an interesting and beautiful thing to see this couple, more or less retired from the world, completely devoted to lofty study of the heavens.

This would have made Henriette a little bitter because it was exactly what she yearned for, and which she did not have, in particular because she had to take care of her mother-in-law when Jules was traveling:

> Mme Huggins is lucky in being able to devote herself exclusively to her husband. Conjugal affection that is not distracted by any other is definitely the one that gives the least deceptions if there is great reciprocity: but we are not made for <u>true happiness</u> here below: the legitimate affections that the law of nature imposes upon us, and which sometimes come to upset trouble the smoothness of intimacy are a necessity of our human condition. No one more than me has experienced the effects of the overriding necessity of sharing oneself between people to whom one owes a more or less complete duty. So I admire M. and Mme Huggins and find that the happiness and calm that they enjoy greatly compensates, and rewards them for this mutual devotion. [23]

When he gave a lecture, Janssen did not hesitate to illustrate his points by carrying out demonstrations, as he did, for example, at the Sorbonne in 1870 to explain the reason for his discovery in 1868:

> Here is an IMAGE OF THE SUN produced by the magic lantern. [...] I place behind this image a screen pierced by a slit. [...] Let us examine the slit with a prism [...] which spreads the white light from the disk into a spectrum. [...] The average intensity of each point of the spectrum is about 5000 times less than direct sunlight. And what about the prominence? [...] The average amount of weakening will be only 5 times. [24]

Conferences on Photography

In 1887, Janssen finished his lecture on celestial photography in Toulouse with these words addressed to young scientists:

> Boldly seize this French discovery, this fruit of the national genius, which we have neglected too much until now. Apply it to all the fields that you cultivate, and you will soon garner legitimate honor for yourselves, beautiful conquests for science, and new glory for our dear country. [25]

That year, the famous international conference on celestial photography took place in Paris. The Académie des sciences sponsored the conference, and its President, Janssen, welcomed the participants:

> The mapping of the heavens [...] where the bodies themselves record their location and their brightness, will form an unprecedented moment in Science. [...] You have sanctioned a new method, and you have established a revolution that will not be any less fruitful than the one that saw the introduction of telescopes into Astronomy. [26]

Although Janssen appears on the group portrait of the conference participants, immortalized by Nadar, he was not directly involved in the great Carte du Ciel enterprise. He was aiming still higher and greater, and the year 1889 was to bring him full satisfaction.

First, he was President of the organizing committee for the International Photographic Conference which was held in Paris in 1889. The program for the meeting was hardly slim, because it was necessary to tidy up matters because of "the extent and multiplicity of branches to which the art of photography has given birth." In short, "it is a question of equipping it with its scientific basis, of unifying its methods and instruments, and of fixing its language" [27], not forgetting its relationship with art. As far as language was concerned, the roots and their usage were specified: for example, "heliophotography" would signify "methods used to obtain photographs of the Sun," whereas "photoheliography" would refer to "photographs for which sunlight is the basis and the agent." Units were defined: those for luminous sources as well as for the sensitivity of emulsions. The meeting also showed that "photography may be a work of art," for which Janssen proposed setting up a school, where students would first study drawing and painting. Some days later, on the occasion of the celebration of the fiftieth anniversary of the discovery of photography, Janssen, the patriot, did not hesitate to propose a toast "to the memory of Niepce and Daguerre, to France, to its genius and to its continuing role at the head of civilization" [28].

A short time afterwards, from 20 to 24 September, at Paris and then at Meudon the last day, Janssen presided over the Conference on Celestial Photography, which did include every specialty: Sun, Moon, planets, meteors, comets, star clusters, stellar spectra, stellar photometry, and nebulae. The conference was due to discuss the methods and instruments best suited to each type of research, but it "will, above all, formulate the arrangements that will be best suited to ensure the conservation of the plates and their reproduction" [29].

Janssen's commitment to the Société française de photographie was particularly active. Janssen became its President for the first time in December 1890. When he passed the seat to Marey 3 years later, the results of his activities are significant: organization of the second international conference, held at Brussels in 1891, during which Pector had proposed the creation of an international union, the basis for which had been drawn up, and the first session of which was held at Anvers in 1892 and the second at Geneva in 1893; and the creation in 1892 yet again of the National Union of photographic societies in France, whose first session took place at Le Havre in 1893.

Until his death, Janssen remained President of the National Union, "a fraternal union, and not a work of centralization and authority," which met at Caen in 1894, at Lyon in 1895 (where Janssen invited ladies "to play their part in the work of photography" [30] and where Louis Lumière filmed Janssen's arrival [31]), at Lille in 1896, and at Nancy in 1898. On 5 January 1900, he again took his place as President of the society, succeeding Lippmann.

The Société Astronomique de France and the Société de Géographie

> From his joining our ranks to his death [...], that is for sixteen consecutive years, Janssen never ceased to show the most active and the most indefatigable commitment to the Société astronomique de France, not sparing his time, nor his strength on its behalf; communicating to us all his work and those of his collaborators, both at the Meudon Observatory and at the Mont-Blanc Observatory. [32]

This is how Flammarion expressed himself when he paid tribute to his friend Janssen in 1920. The two men, mutually appreciative of one another [33], continued to grow close over the course of the years, probably beginning in 1873, when Flammarion was interested in the revolver, then with the creation of Meudon. Previously, Janssen had contented himself with sending notes of his discoveries to "the author of works on astronomy" [34], just as he sent them in the same capacity to the popularizers Amédée Guillemin or Wilfrid de Fonvielle. It is rather strange that Janssen did not join the Société astronomique de France (SAF) before 1891. He is actually only the 453rd member registered! Why did he not accompany his friends and sponsors, dom Pedro II, Bischoffsheim, or Prince Albert 1st of Monaco? Nevertheless 1891 was a year of good recruits for the SAF: Faye, Prince Roland, Eugène Antoniadi, Sir William Christie, Alfred Cornu, Deslandres, Anatole France, Edward Holden (Director of Lick Observatory), and Giovanni Schiaparelli (Director of the Milan Observatory), among others.

Flammarion was delighted, and on 1 January 1892 he "presents to Monsieur, Madame and Mademoiselle Janssen his best wishes for the New Year, and begs the learned Director of the Observatory of Meudon to accept his particular thanks for the honor that he has done to the young astronomical Society by entering into its ranks. We all hope that at one of the shortly forthcoming meetings we shall hear a fine report on physical astronomy" [35].

Fig. 14.1 *Janssen and the Mariani wine*. Mariani Album. © Author's collection

And because Janssen did not belong to a society just in a token role, from 1895 he was President of the SAF for 2 years, and he became a founder member in 1897, which means that he gave it a donation of 500 F. That was the same year that the prize that he founded was awarded for the first time … to Flammarion. In 1902, Janssen actively participated in the organization of the inaugural ceremony on the installation of the Foucault pendulum at the Panthéon, which was a triumph for the SAF. Similar tremendous successes were the festivals of the Sun, organized on the summer solstice by Flammarion and Eiffel, who lent his tower for the event from 1904 onwards. Honorary President of the SAF from 1903, Janssen presided on four occasions. Like many of their contemporaries, the two companions Janssen and Flammarion ("The telescopes of friendship often produce anamorphoses" [36], Jules once wrote to Camille), did not have to be begged to praise the qualities of the "wine of the Sun," the liqueur Mariani (Fig. 14.1).

Janssen was also part of the Société de géographie, where he was again in familiar surroundings. He was, in particular, very proud of having welcomed Prince Roland Bonaparte, who was not only a sponsor, but also a great traveler, a great photographer, and an eminent botanist. And it was he who had the honor and the pleasure, as President of the Society, to welcome another explorer, Prince Henri d'Orléans, in 1896.

Inaugurations, Prize-Givings, Banquets, and Funerals

Janssen's membership of numerous organizations, particularly the Académie des sciences, and his talent as a speaker, meant that it often fell to him to take part in various ceremonies.

From 1877 to 1904, from the death of Le Verrier of whom he was able to praise the "natural strength of character" [37] and the merits (before wondering whether, now that he was in his tomb, gravitation would not completely give way to light), until that of his dear Octave Callandreau, an astronomer at the Paris Observatory, whose passing at the age of 52, he mourned with genuine emotion [38], there were no less than 18 addresses that Janssen gave on the occasion of the passing of outstanding people. It was with the same talent that he retraced the careers of the astronomers Louis Thollon (who had drawn up a visual solar spectrum of 3,000 lines, using a spectroscope of his own construction), Jean-Charles Houzeau, Félix Tisserand, Hervé Faye, and Prosper Henry; the doctors Alfred Vulpian and Léon Gosselin; the chemist Henri Debray; the professor of agricultural chemistry Jean-Baptiste Boussingault; the agronomist Hervé Mangon; the geographer General François Perrier; the engineer Pierre Berthot, so dear to him; the aeronaut Gaston Tissandier; or the "artists" (in the Bureau des Longitudes' interpretation), the manufacturers Émile Brunner and Louis Bréguet. When, at the national civil funeral of Paul Bert, the physiologist doctor and member of the Académie des sciences, thanks to whom aeronauts supplied themselves with bottles of oxygen, but also the man whose life came to an end at the age of 53 at Hanoi, his words inevitably evoke some of the finest pages in any anthology of French literature:

> This coffin thus shows us the reality of the news that, a few week ago, struck the whole of France with amazement: Paul Bert is dead! Paul Bert is dead, which means that France has lost, at the height of his powers, one of its most courageous servants; Science one of its most eminent workers. Paul Bert is dead, which means that the so promising undertaking, where, for the first time perhaps, Science would have become the collaborator of Politics in the work of civilization, is now laid waste! [39]

Paul Bert (Fig. 14.2), who had just taken up his new post as "Résident général *[General Resident]*" for Annam and Tonkin, indeed wanted "to gather round him a galaxy of young scientists, to establish academies, have the country explored scientifically, and show the superiority of our sciences to the learned people in the Orient, who hold intellectual culture in such high esteem; in a word, to follow military conquest with a second, purely peaceful and moral conquest" [40].

As we have seen with that of Saussure, Janssen was also involved in the unveiling of statues. A particularly significant moment for him came in 1879 because he represented the Académie des sciences at Perpignan, where the statue of François Arago was unveiled on 21 September:

> I have just given my speech [41], which, without bragging, I may say was a great success. The platform was in the open air; I had to shout. But I had just had a nice lunch with the Prefect and I was weighed down. My voice was not hoarse. Nothing was heard of the earliest speeches. But the Minister *[Jules Ferry]* (Fig. 14.3) spoke with remarkable vigor. I was frequently applauded at impassioned passages.

Fig. 14.2 *Paul Bert. Copper engraving*. © Author's collection

Fig. 14.3 *Jules Ferry. Félix Potin Album*. © Author's collection

After having spoken, the Minister, the Prefect, and all the Aragos congratulated me. With the one by Paul Bert, which was a virulent political harangue, it was the one that was most successful. Alfred Arago, the senator, gave thanks afterwards. He was crying with emotion. He mentioned me in his thanks and said that he could not tackle the scientific side because genius could not be transferred. It was good for us. Bert said that I was Arago's successor and the best suited to give his eulogy.

Subsequently, we had some delightful music from the mountains of Estagel, melodies that brought tears to our eyes.

Overall, a magnificent ceremony. Enthusiastic populace. Immense crowd in the town. I shall be dining this evening at the town hall. 500 places? It will be long. [42]

The next day, after the dinner, Janssen added: "The Minister speaks very well. An uncommonly vigorous eloquence. The Maire thanked the illustrious scientists who had come from such distances to offer their homage to François Arago" [43]. Before leaving for Estagel, Arago's birthplace, where the proprietor of the sole hotel refused to allow him to pay for his room when he learned that he was one of the gentlemen who had given a speech at the statue, Janssen made "a delightful excursion to the Spanish frontier." He was in good company: the Prefect, the Aragos, and the minister:

During this very intimate trip, I talked a lot to the Minister, and it will not be without benefit for Meudon. He promised, spontaneously, to come and see me. On the way, I was asked about the beauties of the landscape and about those that I had seen on my travels. That is why, later, M. Ferry when speaking of me to an authority said "Here is M. Janssen, an artist and a great scientist". [...] We took a trip on the sea, and one of the fishermen who were taking us, sang us local poems, standing at the stern of the boat, and accompanying it with gestures. We were then sailing gently across the sea, the sunset sky was flushed with pink tints and the silhouette of the singer appeared in bluish tones, which formed a striking contrast. A poor fisherman singing of his loves, whose only theatre was a patch of rocks, unknown to the rest of the world; a minister, thrown into the arena of *[political]* parties and on whom the eyes of the world are fixed; this man whose ideas have no horizon other than his fishing, and the man who has to probe the problems of the Universe – what a contrast to dream about. One might be tempted to ask whose fate was the best. I do not know, but what I do know is that everyone should follow their vocation, thus ensuring the harmony of the Universe. The next day I accompanied the Minister to Narbonne, where I saw the Roman antiquities. [44]

Janssen later speaks of Estagel, where, introduced to the general populace, he had to improvise a speech:

I spoke of the emotion I felt when I saw the feeling here that surrounded the family of the great scientist, that I had just made something akin to a pious pilgrimage by visiting the house that had seen the birth of a scientist that I had taken as a model, that if a son could not visit the family home without emotion, a scientist could not visit the house that saw the birth of the master that had preceded him in his career without a similar emotion. [45]

After Estagel, Janssen returned to Perpignan, where he received some advice:

The Prefect, my host, advised me to go to Barcelona, which is very close, where I would find a rare library of old books originating from the Pillage of the convents. I have been here since Sunday, and during the whole week I have been working in the library, which contains 120 to 150 thousand volumes of rare works, primarily theological and historical, but also works of the science of antiquity. I studied 2 editions of Archimedes, Greek and Latin, which I was extremely happy to skim through, and from which I have taken notes. A Euclid – An Apollonius, very curious works by Kepler, Copernicus. – Galileo's "Nuncius Siderus" with a response by Kepler. Incunabula Bibles, etc. One often makes long journeys to consult such works, which is why I thought I should profit by being in the neighborhood. It costs 25 F to go from Perpignan to Barcelona – I have found accommodation in a Spanish hotel and have been living as a hermit, that is to say as I like. In the morning, I have been working in my room at revising my speech. After lunch I have been going to the Library. Then to the cathedral, where I sat down to draw and to think. I went for a few hikes on a hill from which one looks over the town, of which I have drawn a panorama. Strolling in the town, I found a 16th-century Bible with pictures in the text, which is rare. It was examined by the office of the Inquisition, and suspect passages have been crossed-out, with the signature of the secretary charged with that task. Date 1584. It is a find. [46]

At Estagel, Janssen's speech had of course been mentioned in the papers, as was to be the speech given in 1886 at the Muséum on the occasion of the centenary of the great chemist Eugène Chevreul. The honor given to Janssen was also recounted, supported by an engraving, on the occasion of the great banquet on the day of the anniversary:

As eight o'clock sounded from the clock on the Hôtel de Ville, M. Janssen rose, and said, in the midst of a religious silence: "One hundred years ago, minute for minute, M. Chevreul came into being." [47] (Fig. 14.4)

Janssen also had the pleasure of presenting various prizes, in particular the Janssen Prize of the Académie des sciences, which he founded in 1886, to reward "the person, who in France or from abroad (excluding members of the Institut) shall be the author of a work or a discovery leading to direct progress in physical Astronomy."[6] Between 1892 and 1906, the recipients were Pietro Tacchini, Samuel Langley, George Hale, Henri Deslandres, Aristarch Belopolsky, Edward Barnard, Fernand Foureau, Aymar de La Baume Pluvinel, Alexis Hansky, Annibale Ricco …

Publications

Janssen did not write any original work, and this has certainly contributed to his neglect during the twentieth century. His scientific publications are in the form of reports, certainly many of them, but rather short, published either by the Académie des sciences, or in the journals of the learned societies before whom he had given papers. He undoubtedly treated the Notices edited for the *Annuaires du Bureau des Longitudes* more especially, which he distributed widely as off-prints (which carried a small, personalized label stuck on the covers), and which fitted well with the idea that he held of the scientific author:

A scientific author's integrity comes from their competence, but it is necessary to strip away the technical language in favor of the one that is familiar to the reader. [48]

We have seen how he liked to improvise his words, and that obviously annoyed him when it came to editing detailed technical or scientific reports after the event, in particular after the creation of Meudon.

[6]The corresponding donation in favor of the Institut is the subject of a notarized deed of 26 December 1886, and the definitive acceptance dates from 6 June 1887, AN, MC/ET/CVII/1977.

Fig. 14.4 *The banquet at the centenary of Chevreul, M. Janssen's toast.* L'Illustration, *4 September 1886.* © Author's collection

The Meudon *Annales* arrived extremely late, only in 1896. We may wonder if Janssen published them solely because of the "Trouvelot Affair" [49], the press campaign of hatred unleashed against him by the regrettably famous, anti-Semitic journal edited by Édouard Drumont, *La Libre Parole*, after the death of Trouvelot. The daily accused him, in particular, of having used for traveling expenses in Africa or on Mont Blanc, monies specifically devoted to publications in Meudon's budget. "For seventeen years, the Observatory has not published any report, despite the annual credit of 10,000 F set aside for its publications" [50]. Janssen then explained that he had had to use that money to offset the inadequate sums destined for the buildings and the major instruments.

Relationships with Fellow Astronomers and Colleagues

In company with fellow scientists, Janssen felt remarkably at ease. This was particularly the case with the Académie des sciences, of which he spoke in 1888, when he welcomed his friend Pierre-Paul Dehérain:

> You will see that the Académie is basically a family, not only from the intellectual point of view, but also from the bonds of affection that one forms. At the Académie, one adopts the habits of benevolence and urbanity towards one's fellows of the same age, of deference and respect towards older ones. Within our society that is so active, so militant, and sometimes so merciless towards the weak, or towards those who have no call save on former services, it is good to find oneself in a company that takes account of the past, where age is honored and respected, and where one can calmly regard the evening of one's advancing years and grow old without dread. [51]

Apart from academic tasks, it would seem, in contrast, that true scientific collaboration was non-existent, with one exception: Édouard Stephan, with whom he got along very well.

In 1879, accompanied by his joiner Coffin, Janssen went to observe the partial eclipse of the Sun of 19 July at the Marseille Observatory, where the Director made him very welcome:

> M. Stephan was waiting for us at the station. I had telephoned to him from the station in Paris on leaving. The weather appeared to be getting better. I think, therefore, that we shall be able to observe the phenomenon.
>
> Yesterday I dined with Stephan's family at home. Stephan has three children who are very nice. The youngest is just a few months old. The Director's house is very fine: it was begun by Leverrier *[sic]*, who liked luxury. The observatory is now in a very satisfactory state: a large meridian refractor has been installed. A fully equipped 9-inch equatorial has been added to the 80-centimetre reflector, which dates from Leverrier's time, and with which I worked for a time in 1867, when returning from Greece and Sicily. The garden area and the surroundings are amazing. The town has done a lot for the observatory. It has surrounded it by beautiful high walls, has bought more ground, established avenues, laid on gas, etc. At the main entrance there is a square, which has been called after Leverrier, quite justifiably, because it was he that created the observatory.
>
> [...] We set ourselves up in a small property belonging to, and alongside, the observatory where we are well placed. There is a chemistry laboratory which has made us an excellent photographic laboratory. My 8-inch equatorial from Japan is mounted and ready. [52]

Janssen was so satisfied with his stay that in his report to the Académie des sciences, he expressed the wish that the budget for the Marseille Observatory "currently too modest, should be significantly increased, and that M. Stephan should be able to acquire personnel that would allow the complete utilization of the instruments and of the favorable conditions given by this southern sky" [53]!

Nothing was to be seen of the attitude Janssen had 8 years earlier towards the "young, learned lecturer at the École polytechnique" [54] Alfred Cornu [55] who, like him, was interested in the constitution of the Sun, and with whom he carried on a rather condescending dialogue through the *Comptes rendus de l'Académie des sciences* in 1871: "I have given M. Cornu some friendly objections on the conclusions, rather hasty ones, in my opinion, that he has drawn from his experiments on metallic vapors [...]" [56] ... And although Janssen did not miss the lecture by Cornu at Bordeaux the following year ("M. Cornu is due to give an evening lecture on Monday on the Sun. I shall remain until then, after which I shall fly to join you" [57], he wrote to Henriette), it does not seem that he spoke to him very much.

At Meudon in 1877, the personnel consisted of just three people, Janssen, the photographer Arents, and a caretaker, whose annual salaries amounted to 10,000, 4,000 and 1,200 F, respectively. This rose to a dozen by 1882, at which date the first titular astronomer, Trouvelot was recruited.

If Pasteur replaced Arents in 1880, it was because the latter set himself up as a printer in Paris, and Janssen sometimes had him do work, because he was the one who carried out the heliography of the dirigible *La France*. Otherwise, the personnel remained stable. But they were only assistants: administrative or laboratory assistants, joiners, and caretakers.

Until his death in 1895, Trouvelot remained the sole astronomer other than Janssen. However, after the observation of the eclipse of 1883 from Caroline Island, the relationship between the two men rapidly deteriorated. And yet the work required of Trouvelot was hardly restrictive: "It was agreed that M. Trouvelot would, every sunny day, draw the prominences with their shape, their sizes, and their exact positions on the limb of the Sun. With regard to these observations, it was, in addition, agreed that once this work had been carried out, M. Trouvelot would be at liberty to carry out his personal work. For these various tasks, I entrusted M. Trouvelot with an equatorial with an aperture of 8 in. (0 m 21) which was originally from the expedition that I led to Japan, an equatorial whose refractor was the finest in the Observatory" [58], Janssen explained in the report that he was obliged to send to the ministry after the press campaign which intimated that Trouvelot, who, in addition, had accommodation provided by the job, was "a martyr to science," driven to death by his Director. Janssen's conclusions were clear:

> M. Trouvelot remained at the Observatory for ten years, with no duties. The work for which he was engaged, required a few hours of observation during the day and when the Sun was visible. This daytime duty, could definitely not have affected his health.
>
> [...] For a long time, ever since he joined the observatory, M. Trouvelot was unwell; [...] M. Trouvelot had real merits as an observer, but he had, at the same time, an extraordinary idea of those merits and did not feel that his reputation lived up to those merits. This produced a resentful character, which caused him to complain bitterly. [...] One could therefore excuse him, and I was perfectly happy to do so. But I find that his memory and the interests of his heirs are poorly served by attacks that are so violent, so unjust, so manifestly false, and so easy to repeat. All this does is to compromise the interests that one pretends to serve.

The three other appointed researchers in Janssen's time did not arrive at Meudon until 1897, for Deslandres and Millochau, and 1899 for the youngest (who was just 15!),

Lucien d'Azambuja, who worked only with Deslandres. Between Janssen and Deslandres, the graduate of the Polytechnique who was destined to succeed him, there was far from being a cordial relationship, as we shall see later. On the other hand, Janssen got on well with Millochau, who asked to leave Deslandres' spectroscopy team in 1903 to work with him, and who left again for Paris in 1907. Millochau gave an explanation to the minister at that time:

> Having followed M. Deslandres to Meudon in 1897, on condition that he would have me nominated as assistant astronomer, around 1900 he changed his attitude towards me, without any apparent reasons, and by his arrogance and humiliations made my existence even more difficult by the fact that the young, ignorant people that he employed in his laboratory, followed his example and failed to show me even the simplest respect. *[... In December 1903]*, Monsieur Janssen withdrew me from serving with M. Deslandres and enabled me to work in a different fashion, because it was from that date that my work was published, under my own name, in the Comptes Rendus de l'Académie des Sciences. [59]

The other researchers who worked at Meudon and at Mont Blanc, with or for Janssen, were volunteers. Although another small "affair" arose with Louis Rabourdin, who photographed nebulae and star clusters with the 1-m telescope between 1897 and 1899, and who would not return until 1908 under Deslandres ("I have the honor to inform you that M. Janssen has withdrawn in a decisive manner my usage of the telescope that I have been using since the beginning of 1897. It is thus impossible for me to continue the research that I have undertaken and which have already given me some curious results" [60], he wrote to the Director of Higher Education in June 1899), relations were excellent with the others. These were the Serb Georg Stanoïevitch, the Russians Hansky and Tikhov, the Slovak Milan Stefanik, introduced to Janssen by Flammarion in 1904, and Count Aymar de La Baume Pluvinel. The last was particularly close to Janssen. Greatly moved at the time of his death, he wrote to Flammarion from his property in the country (in the department of Cantal):

> When I worked with him, already some fifteen years ago, I was subjected to the effects of his infectious energy, and I returned from my visits to Meudon with my head full of new ideas and with a new enthusiasm for work. [61]

References

1. Janssen, Jules, "Conférence sur l'éclipse totale du 18 août 1868", in *Lectures académiques discours*, 299–329, p. 300.
2. Janssen, Jules, Bulletin de la Société philomathique de Paris, 8th series, **I** (1888–89), 10.
3. Administrative correspondence of the Meudon Observatory. Paris Observatory Library.
4. Janssen, Jules, "Allocution au dîner offert à M. Simon Newcomb le 29 mai 1903", in *Œuvres Scientifiques*, Vol. II, p.593.
5. Letter from Jules Janssen to Norman Lockyer, 01 December 1873, EUL.
6. Letter from Norman Lockyer to Jules Janssen, 30 December 1873, BIF Ms 4133-358.
7. Lockyer, J. Norman, *Contributions to Solar Physics*, London, Macmillan and Co., 1874.
8. Letter from Jules to Henriette Janssen, 22 September 1873, BIF Ms 4133–176.
9. Letter from Jules to Henriette Janssen, 25 September 1873, BIF Ms 4133–171.

10. Letter from Jules to Henriette Janssen, 28 September 1873, BIF Ms 4133–178.
11. Letter from Jules to Henriette Janssen, 10 October 1873, BIF Ms 4133–180.
12. Letter from Jules to Henriette Janssen, 16 August 1878, BIF Ms 4133–205.
13. Letter from Jules to Henriette Janssen, n.d. *[20 August 1878]*, BIF Ms 4133–204.
14. Letter from Jules to Henriette Janssen, n.d. *[24 August 1879]*, BIF Ms 4133–209.
15. An interrupted draft of a letter from Jules Janssen to William Huggins, n.d. *[1879]*, BIF Ms 4135–329.
16. Letter from Jules to Henriette Janssen, 16 September 1888, BIF Ms 4133–290.
17. Fauque, Danièle, "Observations astronomiques français et diffusion de l'astronomie à l'Association française pour l'avancement des sciences (1872–1914)", in *Observatoires et patrimoine astronomique français*, ed. G. Boistel, Lyon, SFHST/ENS-LSHE Éditions, 2005, 175–98.
18. Letter from Jules to Henriette Janssen, 04 September 1875, BIF Ms 4133–184.
19. Letter from Jules to Henriette Janssen, 13 September 1876, BIF Ms 4133–195.
20. Janssen, Jules, "Conférence sur l'éclipse totale du 18 août 1868", in *Lectures académiques discours*, 299-329, pp. 300–1.
21. Letter from Jules to Henriette Janssen, August 1877, BIF Ms 4133–201.
22. Becker, Barbara J., *Eclectism, opportunism, and the evolution of a new research agenda: William and Margaret Huggins and the origin of astrophysics*, Ph.D. dissertation, The Johns Hopkins University, 1993; Becker, Barbara J., *Unravelling Starlight: William and Margaret Huggins and the Rise of the New Astronomy*, Cambridge (UK), Cambridge University Press, 2011.
23. Letter from Henriette to Jules Janssen, 01 September 1877, BIF Ms 4134–174.
24. Janssen, Jules, "Conférence sur l'éclipse totale du 18 août 1868", in *Lectures académiques discours*, 299-329, pp. 317–18.
25. Janssen, Jules, "La photographie céleste" in *AFAS, Compte rendu de la 16e session, congrès de Toulouse (1887)*, 1888, 395–409.
26. Janssen, Jules, "Allocution adressée aux membres de la Conférence internationale de photographie céleste, à la séance de l'Académie des sciences du 18 avril 1887", *CRAS*, **108** (18 April 1887), 1067–70.
27. Janssen, Jules, "Discours prononcé à la première séance", in *Exposition universelle internationale de 1889. Congrès international de photographie, tenu à Paris, au palais du Trocadéro, le 6 août 1889, Procès-verbaux et résolutions*, p. 5.
28. Janssen, Jules, "Discours prononcé à la fête du cinquantenaire de la divulgation de la photographie, le 19 août 1889", *Bulletin de la Société française de photographie,* **5** (1889), 263–8.
29. Janssen, Jules, "Discours prononcé à la séance d'ouverture du congrès de photographie céleste, tenu à Paris du 20 au 24 Septembre 1889", *Annuaire du BdL pour l'an 1890,* 722–33.
30. Janssen, Jules, "Allocution prononcée le 11 juin 1895", in *Œuvres Scientifiques*, Vol. II, 376–8.
31. The sequence appears in the film *Incunables du cinéma scientifique*, Meudon, CNRS diffusion, 1984.
32. Flammarion, Camille. in *Inauguration de la statue de Jules Janssen,* 31–32, p. 32.
33. Launay, Françoise, "La correspondance Janssen/Flammarion", *l'Astronomie,* **121** (2007), 276–84.
34. Letter from Jules to Henriette Janssen, 02 February 1868, BIF Ms 4133–99.
35. Card from Camille Flammarion to Jules Janssen, 01 January 1892, BIF Ms 4135–167.
36. Letter from Jules Janssen to Camille Flammarion, 03 January 1900, SAF, Flammarion Juvisy collection.
37. Janssen, Jules, "Discours prononcé aux funérailles de Urbain Le Verrier le 25 septembre 1877" in *Lectures Académiques, discours*, 133–9.
38. Janssen, Jules, "Discours prononcé aux funérailles de M. Callandreau le mardi 16 février 1904", in *Œuvres Scientifiques*, Vol. II, 602–4.

39. Janssen, Jules, "Discours prononcé aux funérailles de Paul Bert le 15 janvier 1887", in *Lectures Académiques, discours*, 161–5, p.161.
40. *Ibid.*
41. Janssen, Jules, "Discours prononcé à l'inauguration de la statue de François Arago le 21 september 1879", in *Lectures Académiques, discours*, 141–53.
42. Letter from Jules to Henriette Janssen, 21 September 1879, BIF Ms 4133–214.
43. Letter from Jules to Henriette Janssen, 22 September 1879, BIF Ms 4133–218.
44. Letter from Jules to Henriette Janssen, 29 September 1879, BIF Ms 4133–215.
45. *Ibid.*
46. Letter from Jules to Henriette Janssen, 04 October 1879, BIF Ms 4133–211.
47. *L'Illustration*, No. 2271, 04 September 1886, p.164.
48. Janssen, Jules, "De l'art de répandre les notions scientifiques", speech at the second reunion of contributors to *La Nature*, in *Œuvres Scientifiques*, Vol. II, pp. 82–4.
49. Launay, Françoise, "Trouvelot à Meudon: une "affaire" et huit pastels", *l'Astronomie*, **117** (2003), 452–61.
50. Méry, Gaston, in *La Libre Parole*, 27 May 1895.
51. Janssen, Jules, "Réception de P.P. Dehérain", in *Lectures Académiques, discours*, 263–8, p.267.
52. Letter from Jules to Henriette Janssen, 18 July 1879, BIF Ms 4133–207.
53. Janssen, Jules, "Sur l'éclipse du 19 juillet 1879 observée à Marseille", *CRAS*, **89** (11 August 1879), 340–2.
54. Janssen, Jules, "Sur la constitution du Soleil", *CRAS*, **73** (14 August 1871), 432–6.
55. Le Gars, Stéphane & Maison, Laetitia, "Janssen, Rayet, Cornu: trois parcours exemplaires dans la construction de l'astronomie physique in France (1860–1890)" *Revue d'histoire des sciences*, **59** (2006), 51–81.
56. Janssen, Jules, "Sur ce qu'ont jusqu'à ce jour d'incomplet les résultats d'observations spectroscopiques pour nous faire connaître la constitution du Soleil" *CRAS*, **73** (25 September 1871), 793.
57. Letter from Jules to Henriette Janssen, 07 September 1872, BIF Ms 4133–166.
58. Janssen, Jules, "Rapport au Ministre", 26 June 1895, AN F^{17} 3747.
59. Letter from Gaston Millochau to the minister, 05 March 1907, cited by Véron, Philippe, *Dictionnaire des astronomes français, 1850–1950,* unpublished.
60. Letter from Louis Rabourdin to the Director of Higher Education, 15 June 1899, AN F^{17} 3745.
61. Letter from Aymar de La Baume Pluvinel to Camille Flammarion, 31 December 1907, SAF, Flammarion collection at Juvisy.

Epilogue

"L'homme n'est encore qu'à la préface du livre qu'il est appelé à écrire sur l'Univers."[1]

[As yet, Man is only at the preface of the book that he is called to write on the Universe.]

Jules Janssen

The Last Ascent: Vesuvius in a Sedan-Chair

It was on December 14, 1904, at an age of more than 80, that Janssen decided to "visit" Vesuvius again. He was accompanied by the faithful Millochau and, after having climbed part of the mountain on the funicular, he took a carrying chair with porters to reach the summit. When they arrived, the astronomers were subjected to "a considerable discharge of lapilli and volcanic bombs, which, thrown more than 30 m into the air, fell around them."[2] Janssen only just avoided being hit, because one of these 30-cm bombs landed 4 m away from his chair. Janssen did not consent to the panic-stricken guides' descent until his work was finished: he had photographs taken and samples collected. Because the location chosen was not adequate for examining the changes in sunlight that arose when it had passed through the vapors escaping from the crater, Janssen decided to have himself carried elsewhere the next day, which left time for Millochau to obtain samples of the gas. The operation was obviously not without danger, because the bottles were lowered 10 m down into the crater. On 16 December, the team reached another location with porters. There, a large grating spectroscope, as well as a 108-mm refractor, fitted with a direct-vision spectroscope were quickly installed, and Janssen had some photographs of the solar spectrum taken (which he intended to study later), under different

[1] Jules Janssen's notebook, 30 July 1884, BIF Ms 4129–9.

[2] Janssen, Jules, "Sur une récente ascension au Vésuve", *CRAS*, **140** (23 January 1905), 200–2.

F. Launay, *The Astronomer Jules Janssen: A Globetrotter of Celestial Physics*, Astrophysics and Space Science Library 380, DOI 10.1007/978-1-4614-0697-6, conditions: the Sun behind the fumes escaping from the crater, the Sun beginning to move clear of the fumes, and the Sun away from Vesuvius, as well as some spectra of the diffuse light from the sky.

The Last Eclipse

Being closely occupied by Mont Blanc and his various meetings, Janssen had not left France to travel to observe an eclipse since that of 1883 at Caroline Island. So it was the photographer Pasteur, the "pupils" Stanoïevitch, Stefanik, the Spaniard Landerer, and above all the associate astronomer La Baume Pluvinel, who hardly missed any except Senegal in 1893, who represented Janssen's observatory in eclipse expeditions abroad. In 1900, however, it was the astronomer Deslandres who directed the expedition in Spain.

On August 30, 1905, the duration of the eclipse was close to 4 min, and the line of totality passed through Spain. While Deslandres directed the expedition that had been entrusted to him by the Bureau des longitudes at Burgos, on which he took Kannapell and d'Azambuja, and complained that he had too few instruments at his disposal, Janssen went to Alcosebre, together with the astronomers Millochau and Stefanik, and the photographer and assistant photographer Pasteur and Coroyer.

The aims of the expedition were to "study the progress of the phenomenon by means of photography; the form and details of the corona and the prominences; the spectra of the various gaseous envelopes of the Sun; and the meteorological changes produced by the eclipse,"[3] but Janssen was there only as a spectator.

Pasteur obtained photographs that showed the prominences and the higher level of the chromosphere, and the middle section of the corona. As for the coronal jets, they extended to 1°23.5′ from the Sun's limb, that is to more than 5 solar radii.

In the C line of hydrogen, and the D_3 line of helium, Stefanik observed three fine prominences, but the monochromatic image of the corona was essentially absent. It was, by contrast, extremely brilliant in the green line of "coronium."

As for him, Millochau noticed that the spectrum of the prominences primarily consisted of the lines of hydrogen and helium and that these lines were not coronal, whereas the continuous spectrum of the corona showed a maximum between the D_3 line and the green line.

Stefanik and Millochau prepared several papers for the Académie des sciences, which Janssen presented and commented upon. Contrary to Janssen's normal practice, quantitative results were given in the papers. In addition, these publications were gathered together in a section of the *Annales* where some beautiful heliographic plates were shown.[4]

---

[3]Janssen, Jules, "Observation de l'éclipse totale du 30 août 1905 à Alcosèbre (Espagne)", *CRAS*, **141** (09 October 1905), 569–71.

[4]Janssen, Jules, *Annales de l'Observatoire d'astronomie physique de Paris*, Vol. 3, first Part, Paris, Gauthier-Villars, 1906.

The Last Farce

Janssen would not leave his post as Director because he enjoyed a special dispensation. Deslandres therefore had to wait a long time for his expected succession, following his transfer from Paris to Meudon in November 1897. It was not until July 1906 that he was named Assistant Director in charge of administrative matters, as a result of the intervention of the senator Boudenoot, an ex-student of the Polytechnique. The latter had actually written to the minister for public education on January 12, 1906:

> I intended, when there was discussion of the budget for public education, to ask you a question about the administration and functioning of the Meudon Observatory, which appear to me to be compromised, as well as the interests of science, by the maintenance in his post of a man who is 82 years of age.[5]

On 19 July, the minister, Aristide Briand, wrote to Janssen:

> As I assured you, during the conversation that I recently had with you, I am fully disposed to retain you as head of the Observatory of Meudon until the Conference that is due to take place there in 1907. The numerous and eminent services that you have given to science, your beautiful discoveries in such a new field as physical Astronomy, the actual creation of this Observatory, founded on your initiative, create a duty for me to defer to the desire that you have expressed. I beg you, therefore, to see in this decision a mark of the high esteem of the Government for the scientist whose work has brought honor to our country. But I have made an order, by the terms of which M. Deslandres is made responsible for the duties of Assistant Director, and administrator; I am convinced that he will be a useful and competent collaborator for you, etc.[6]

On July 28, 1906, there took place at Meudon "the installation of Deslandres as Assistant Director," a proper farce, the principal scene of which was the object of a printed report that was sent to the administration in December 1906. All the personnel, with the exception of Millochau and the caretaker Guillot, who were on holiday, were assembled in the library. The report was written by the mechanic Chevalier, the book-keeper Heymann, the photographer Pasteur, and Janssen. The extracts that follow give an idea of the atmosphere that prevailed at Meudon that day:

Janssen

> I have come to this meeting to tell you that I remain as Director here, but that the Minister has charged M. Deslandres with the administration of the Observatory. You must, therefore, obey him in future [...]
>
> *M. Deslandres rushes into the room and, without taking off his panama hat, shouts in fury at M. Janssen:*

Deslandres

> I have given the order for the personnel to be called individually, and then you, you call them together collectively! This must stop! You are nothing any more! Nothing but honorary

[5]Cited by Véron Philippe, in *Dictionnaire des astronomes français, 1850–1950.*

[6]Letter from Aristide Briand to Jules Janssen, 19 July 1906, a type-written copy, SAF, Flammarion collection at Juvisy.

Director, all the administration belongs to me; I told you that yesterday, but you're splitting hairs, you're splitting hairs ...

Janssen, choking

No, I am not splitting hairs; moreover you owe me some deference. [...]

Deslandres

[...] What are you complaining about? I could have taken everything; I could have had myself named Director, I was offered it, but I didn't want to. [...] I didn't want to accept; all these material advantages that I could have taken, I've left them to you; you owe me all of that and ...

Janssen

Yes, I know what I owe you! Go away!

Deslandres, to the personnel

Well! Well!

the personnel

....

Deslandres, to Janssen

Ah, so it's me?

Janssen

Yes, you!!!

Deslandres

Right, right!

He exits smiling, and goes and hides behind the half-open door.

Janssen

No, I cannot accept this, and I retract what I have said, you will obey the orders of M. Deslandres only after I have confirmed them.[7]

Hostilities had clearly been opened, as the sequel to this event confirmed. As when Stefanik was on an expedition to Mont Blanc in August 1906, Deslandres had the doors to the apartment that he occupied at the observatory forced, so that he could install a family who did not belong to the administration. In the following December, Deslandres dismissed Heymann, a Dutch citizen who had been employed at the observatory since December 1881, and he sent Kannapel away, who for 3 years had been assigned to carry out his own orders.[8] After Janssen's death, the war continued with Mme Janssen and her daughter, who left their dwelling in the observatory on September 28, 1908 to move to the avenue du Château at Bellevue.[9]

---

[7] *Observatoire de Meudon, Installation de M. Deslandres comme Directeur-Adjoint*, 1906, SAF, Flammarion collection at Juvisy.

[8] Véron Philippe, in *Dictionnaire des astronomes français, 1850–1950.*

[9] Down the hill from Meudon.

"Deslandres had the moving vans blocked in, because they contained a certain number of instruments that had belonged to Janssen personally, in particular two reflectors with apertures of 20 and 37 cm, which Deslandres maintained were the property of the observatory. They had to obtain an order from the civil tribunal at Versailles to allow the move to go ahead!"[10]

Twelve years later, did Deslandres really think that he had definitely been involved in "hostility" when he pronounced the following words in front of Janssen's statue?

> It is always difficult to have new ideas accepted; and without it being necessary to stress this, Janssen often came up against indifference or even hostility. He was able to succeed thanks to his exceptional qualities. Very intelligent, even having a lot of wit, he was interested in all the great questions and, as he was at the same time a scientist, an artist, and a poet, he knew how to express his ideas and his projects with particular charm. His way of speaking was exquisite, and his powers of persuasion were very great. The natural gifts greatly served the cause of Astronomy and the Observatory of Meudon.[11]

The Last Family Photo

It was following a suggestion by the American, George Hale, that an international association for solar studies should be created that a steering committee met at Oxford in September 1905, after the eclipse. France was represented by Deslandres, La Baume Pluvinel, Alfred Perot of the CNAM, Charles Fabry then of the Faculty of Sciences at Marseille, and Janssen, who went to Oxford with Henriette and Antoinette. These ladies are in the group photograph, where Lockyer is seated next to Janssen, the honorary President. In France, the daily press reported the event in an article that dealt with prediction of the weather: "The immortal work of M. Janssen and those of the New solar Union have, in reality, the same end *[the improvement of the laws governing atmospheric phenomena]*, as well as studying the constitution of the body that illuminates us."[12]

Two years later, in May 1907, the young association met at Meudon, and it was there that it decided "to adopt the value of 6,438.4696 units of Angström for the wavelength of the red line of cadmium. The unit thus defined coincided, within the accuracy available at the time, to one ten-millionth of a millimeter."[13] Although at Oxford there were only 33 people in the photograph, including five ladies, there was more than double the number at Meudon (Fig. Epi. 1). Retained as Director thanks to this conference, Janssen, the patriarch, "immutably dressed in a black frock coat and wearing a large white cravat,"[14] is seated this time between his wife and daughter; but friend Lockyer is not far away, between Mme Janssen and Weiss.

---

[10] *Ibid.*

[11] Deslandres, Henri, in *Inauguration de la statue de Jules Janssen*, 15–23, p. 22.

[12] *Le Petit Journal*, 12 October 1905, p. 2.

[13] Salet, Pierre, *Spectroscopie astronomique*, Paris, Doin, 1909, p. 145.

[14] *Le Petit Journal*, 24 December 1907, p. 1.

Fig. Epi.1 *The Conference of the Union for Co-operation in Solar Research (Meudon, May 1907). Seated from left to right: Father Cirera, Weiss, Lockyer, Mme Janssen, Janssen, Melle Janssen, ?, Deslandres. Hale is behind Weiss.* © Paris Observatory Library

December 23, 1907

On the day that Janssen's statue was unveiled, Flammarion finished his speech with these words:

> On his deathbed, two hours after he drew his last breath, I admired that fine face, noble and quiet; he seemed to be asleep. He was 83 years old, which is not a great age for astronomical longevity.[15]

As was customary at the period, photographs were taken showing Janssen on his deathbed. In the villa des Brillants at Meudon, there is also the mortuary photograph of Rodin, Janssen's neighbor, of whom it is difficult to believe that they had not occasionally met. They were, at least, both present at the Panthéon at the major celebration for the inauguration of the Foucault pendulum.[16]

"A slight cold, which degenerated into congestion," had carried off Janssen "with no suffering"[17] on December 23, 1907, at half-past-four. Janssen thus departed on the day of the winter solstice; it was also the day of the funeral of Lord Kelvin, his friend Sir William Thomson.

---

[15] Flammarion, Camille, in *Inauguration de la statue de Jules Janssen*, 31–3, p. 32.

[16] La Cotardière, Philippe de, and Fuentes, Patrick, *Camille Flammarion*, Paris, Flammarion, 1994.

[17] Bigourdan, Guillaume, "J. Janssen", *Bulletin astronomique*, **25** (1908) 49–58.

Fig. Epi.2 *Scientia Conference dinner menu dedicated to Janssen in 1891.* © Audouin Dollfus collection

Passing into the Land of Memory

On December 24, 1891, the 17th dinner of the Conférence Scientia was dedicated to Janssen. In his response to the speech by Gaston Tissandier in his honor, Janssen pronounced one of the beautiful sentences of which he was fond:

> What more gentle reward, what demonstration that could go more directly to the heart of a studious man *[could there be]* than those marks of warmth and of esteem given by our peers, by those that follow the same career, devote themselves to the same studies, know the same joys and the same bitterness, and who are the most appropriate and unimpeachable judges to appreciate our work and our life? [...] This is a coin that may not consist of brilliant and resounding metal, but which is none the less solid, rare and precious. So, gentlemen, I am persuaded that, after our death, when it comes to crossing the river of forgetfulness,

Fig. Epi.3 *Jules Janssen in 1903. Copper engraving by Achille Jacquet.* © Author's collection

> and to pay the ferryman who guides the boat leading to the land of memory, this is the only coin that will guarantee our passage.[18]

There can be no doubt that he was accorded passage to the land of memory without difficulty in 1907. Thank you, reader, for making sure that it remains there as long as human memory persists.

[18] Janssen, Jules, "Discours prononcé au dix-septième dîner de la Conférence Scientia le 24 décembre 1891", in *Œuvres Scientifiques*, Vol. II, 253–9, p. 255.

Chronology

(*: astronomical observation or result, ☐: invention, □: observation of solar eclipse)

1824	22 February: birth of Pierre *Jules* César Janssen in Paris
1824	23 February: baptism at église Saint-Roch
1828	Birth of Philiberte *Henriette* Forestier in Paris
1832	Janssen is the victim of an accident (a fall) which leaves him lame
1835	Lives in rue des Vieux Augustins in Paris
1836	Lives (in Paris): rue de Londres, then rue de Bellefond, and then rue des Bons-Enfants
1836	First communion
1840	12 October: lives as 9, rue Rochechouart, in Paris
1840	12 October: enters the Banque Tharaud
1842–1844	Lives (in Paris), in rue Royale, at no. 10, and then at no. 14
1847	30 September: leaves the Banque Tharaud
1847	1 October: joins M. Boulet
1848	July: joins M. Lapelle
1849	13 January: baccalaureate of letters
1850	22 November: baccalaureate of mathematical sciences
1851	August: travels to London
1851	Follows courses at the Sorbonne
1852	31 July: licenciate in mathematical sciences
1853	January: supply teacher at Lycée Charlemagne; contacts with Le Verrier
1853	August-September: travels to Switzerland
1854	18 October: employed as tutor by M. Grandidier at Fleury-Mérogis
1855	24 November: licenciate in physical sciences
1856	Travels to Constantinople, Rhodes, Cyprus, Jerusalem, Alexandria, Cairo, Malta
1856	18 August: visits his uncle at Roanne
1857–1858	Voyage to South America with the Grandidier brothers
1859	1 January: proposes to Henriette Forestier
1859	From 1st April to 13 August: tutor to Henri Schneider, at Le Creusot
1859	19 September: marries Henriette Forestier in Paris

(continued)

(continued)

1859	21 September: religious marriage at église Saint-Merry
1859	21 September: lives at 87, chaussée Clignancourt in Paris
1860	9 May: leaves again for Le Creusot
1860	25 May: death in Paris of his father Antoine César
1860	11 August: a girl is still-born
1860	17 August: doctorate in sciences
1861₪	Constructs an ophthalmoscope with Dr. Follin, an ophthalmologist
1861	14 March: moves to 21 rue Labat, in Montmartre, with his wife and mother
1861	7 November: birth of his daughter Antoinette in Paris
1861	4 December: death of Mme Forestier, Henriette's mother
1862₪	Devises several spectroscopes which he has built by Hofmann
1862*	19 October: expedition to Italy until April 25, 1863: Rome, Turin, Genoa, Florence
1862*	Demonstrates the terrestrial origin of dark irregular lines that he observes in the solar spectrum; proposes that these should be called "*telluric lines*"
1864*	Strasbourg, Geneva (first observations at altitude), Puy de Dôme
1865	August: Brittany
1866	Travels to England: London and Nottingham
1866	Receives part of the 1865 Bordin prize from the Académie des sciences
1866*	Studies the spectrum of water vapor in the laboratory (at the gasworks at La Villette)
1867□	March: annular eclipse at Trani (Italy)
1867	April: travels to Greece and to the island of Santorini
1867	Devises the "brush thermometer"
1867*	May: travels to Palermo and to the summit of Etna; announces the presence of water vapor in the atmospheres of Mars and Saturn
1867	20 July: Lisbon
1867*	August, October: expedition to Spain: Madrid, Toledo, Burgos, Cordova
1868	7–12 June: London
1868	16 June: leaves for British India
1868□	19 August: the day after the major total eclipse of the Sun that he had just observed at Guntoor, he demonstrated the possibility of observing the spectral lines of solar prominences outside eclipses
1868	28 October: appointed chevalier de la Légion d'honneur
1868*	December: spends 5 months in the Himalaya
1869	Receives the Lalande prize (increased fivefold) from the Académie des sciences
1869₪	Describes the principle of the spectroheliоscope which enables monochromatic images of luminous bodies to be obtained
1869	Visits Pierrefonds with Viollet-le-Duc
1869	August: attends BAAS meeting at Exeter, with Henriette
1869	December: important meeting at the Ministry for Education
1869–1870	visits Victor Duruy; transports skulls to Quatrefages
1870	22 May: sees Delaunay: The Pavillon de Breteuil at Sèvres is placed at his disposal, and he stays there until 30 August
1870	17 July: in England

(continued)

(continued)

1870	12 September: a girl is still-born
1870	16 November: sees Nadar at Auteuil
1870□₪	2 December: leaves besieged Paris by balloon to go to observe an eclipse in Algeria; during the flight, he invents the aeronautical compass, which determines direction and velocity of aerostats; during the course of the trip, he carries a verbal message to Gambetta at Tours
1871	6 January: death in Paris of his mother, Pauline Marie Lemoyne
1871	January–March: stays at Bordeaux to be near the seat of government
1871	15 May: given a mission to England
1871	17 July: London
1871	Made "Doctor of laws" at Edinburgh
1871□	Returns to India to observe an eclipse, with a double refractor, aiming to study the solar corona
1872	Elected Associate of the Royal Astronomical Society
1872	August: meeting of the BAAS at Brighton
1872	September: meeting at Bordeaux
1873	August: Lyon, conference on the Sun
1873	September–October: Bradford, Newcastle, London, Greenwich where he meets Airy
1873	10 February: elected to the Académie des sciences
1873₪	Conceives his "photographic revolver", the precursor of the cinematographic camera, to observe the transit of Venus in 1874
1873	26 May: elected member of the Bureau des longitudes
1874	August: leaves for Japan
1874*	8 December: observes the transit of Venus from Japan
1875□	From Japan, travels to Siam (now Thaïland) to observe an eclipse
1875	26 June: returns from Siam
1875	24 August–10 September: England (Bristol, Oxford)
1875	Becomes foreign member of the Royal Society
1875	6 September: obtains the decree creating the "Observatoire d'Astronomie Physique de Paris", of which he will be Director until his death
1875	October: stays with Labiche in Sologne
1876	28 May: visits la Malmaison with the architect Moyaux
1876	August: excursion to Royat and to Thiers
1876	13 October: settles in Meudon
1876	6 November: sees the optician Feil at rue Lebrun
1876	2 December: receives the Royal Society's Rumford Medal
1877*	Begins routine photographic monitoring of the Sun at Meudon
1877	8 February: promoted to officer of the Légion d'honneur
1877*	Claims that "photography will soon be the actual retina of the scientist" (this expression is not published until 1882)
1878	26 May: David Gill visits Meudon
1878	16 December: photographs Paris under the snow from his kitchen at Meudon
1879	15 April: the law assigning the Meudon estate to the Observatory of Physical Astronomy is promulgated
1879	May: sits for Henner

(continued)

(continued)

1879□	19 July: observes an eclipse from Marseille
1879₪	Conceives a special shutter (1/6,000 s) for solar photography
1879	21 September: makes a speech at the unveiling of a statue of Arago in Perpignan
1879	October–December: Estagel, Barcelona (visits the Library)
1880	September: takes the cure at Plombières
1881	4 January: Gambetta visits Meudon
1881	February: meeting at Alger
1881*	30 June: he is the first to photograph the whole of a comet
1881₪	Invents the photographic photometer and the method of stellar circles
1881	September: visits the painter Jean-Jacques Henner at Bernwiller
1881	10 December: sits for the sculptor Paul Dubois
1882*	6 December: observes the transit of Venus from Oran (Algeria)
1883□	6 May: observes an eclipse from Caroline Island (Oceania)
1883*	Tahiti, Hawaï; night observations at Kilauea crater
1883	San Francisco, Mormons, Colorado Springs, Chicago, Madison, Washington, Boston, New York
1883	12 September: travels in Germany and Austria (Stuttgart, Munich, Nuremberg, Salzburg)
1883	October: in England with Henriette and Antoinette
1884	28 March: Lockyer visits
1884	4 May: Tacchini visits Meudon
1884	3 August: sits for Henner; sees Redier about the "revolver"
1884	August: excursion to Annecy
1884	September: buys a plot at the Père-Lachaise cemetery
1884	October: leads the French delegation at the "Meridian Conference" at Washington; argues for a neutral meridian; travels to New York and Canada
1885	Decides to equip the Meudon Observatory with two large instruments, a double refractor with a focal length of 16 m, having a visual objective 83 cm in diameter (the largest in Europe and the third largest in the world) and a photographic objective, 62 cm in diameter, together with a reflector with an aperture of 1 m, working at f/3.
1885*	Carries out spectroscopy experiments at Meudon
1886	25 March: attends the marriage of the daughter of Charles Gounod
1886	September: Mont Dore [in the Auvergne]
1886	Has an apartment in Paris, 63 rue de Vaugirard
1887	Attends the Conference on the Carte du Ciel in Paris
1887*	Makes observations at the Pic du Midi
1887–1888	President of the Académie des sciences
1887	30 August: excursion to Le Brévent [a mountain in Haute-Savoie]
1888	29 January: visits the Eiffel Tower
1888*	12–16 October: ascends Grands-Mulets in a chair carried by porters
1888	14 November: inauguration of the Institut Pasteur
1888	3–21 September: BAAS meeting at Bath
1889	23 April: phonograph session at the Académie des sciences
1889	25 April: phonograph session at 63 rue de Vaugirard

(continued)

(continued)

1889	27 April: phonograph session at the Académie des beaux-arts
1889	September: chairman of both Congrès International de photographie and Congrès de photographie céleste
1890*	March–May: spectroscopic work in Algeria (Biskra-Tuggurth, El Oued).
1890	August: Conference at Limoges
1890*	17–22 August: first ascent to the summit of Mont Blanc
1891	26 September: photograph taken by Tairraz from the summit of Le Brévent (a mountain in Haute-Savoie)
1891	31 March: opening of the Carte du Ciel Conference
1891	4 May: sits for Nadar
1891	25 August: photographic conference at Brussels
1892	February: sees the architect Vaudremer about the observatory for Mont Blanc
1892₪	Devises a correcting lens for keratoconus
1892	6 May: receives the medal from the Société française de photographie
1893	19 February: Lockyer dines at Meudon
1893	22 March: funeral of Jules Ferry
1893	4 May: visit by Eiffel
1893	28 June: Camille Saint-Saëns visits the Grande coupole and the telescope
1893	29 July: promoted to Commander of the Légion d'honneur
1893	11 August: Besançon
1893	Photographic conference at Geneva
1893*	8–12 September: second ascent of Mont Blanc where the observatory is opened
1893*	Experiments on Mont Blanc about the laws governing oxygen absorption
1893*	15 December: first observation (of the Pole Star) with the grande Lunette at Meudon
1894*	12 January: first observation of the Moon with the grande Lunette
1894	19 January: visits Jules Simon about the latter's eyes
1894	17 September: dines with Whymper at Chamonix
1894	28 November: funeral of Victor Duruy
1894	26 December: elected President of the Bureau des longitudes
1895*	Algeria, Barcelona, Madrid
1895*	28 September: third ascent to the summit of Mont Blanc
1895	22 October: Lockyer dines at Meudon
1895	27 October: Lord Kelvin dines at Meudon
1896	Roubaix, Rouen, Lille
1896	13 June: funeral of Jules Simon
1896	Chamonix, Annecy
1896	9 November: the Janssen Prize of the Académie des sciences is awarded to Deslandres
1897	January: Mulhouse
1897	June: Reims
1898	December: dines with Nadar
1899	May: conference at Rennes
1900	26 May: lecture at the Palais de l'Optique of the Paris Universal Exhibition
1900	23–28 July: photographic conference (at Meudon on 27th)

(continued)

(continued)

1900	19 October: receives a visit from Deloncle about the Exhibiton's Great Refractor
1902	22 February: goes to the Panthéon for the ceremony marking the centenary of the birth of Victor Hugo
1902	5 May: meets the ophthalmologist Émile Javal, talka about keratoconus
1902	22 October: Foucault pendulum at the Panthéon
1903	17 July: Fécamp
1903	October: Florence and Milan; November: Rome, with Henriette et Antoinette
1903*	Publishes his *Atlas de photographies solaires* consisting of a selection of photographs taken at Meudon between 1876 and 1903, using a photoheliograph that he had devised
1904*	Ascends Vesuvius in a sedan-chair
1905	Receives the "Progress Medal" from the Royal Photographic Society
1905□	Goes to Spain to observe his last eclipse of the Sun
1907	May: presides at the Conference of the International Union for Cooperation for solar research
1907	23 December: decease at Meudon Observatory
1913	1 April: decease at Meudon of Henriette Janssen
1920	31 October: unveiling of his statue at Meudon
1924	27 February: decease of Antoinette Janssen at Meudon

Bibliography

Jules Janssen's Printed Works

Annales de l'Observatoire d'Astronomie physique de Paris, sis Parc de Meudon, Paris, Gauthier-Villars, vol. 1 (1896), vol. 2 (1906), vol. 3 (1st part 1906, 2nd part 1907).

Lectures académiques, discours, Paris, Hachette, 1903.

Œuvres scientifiques, recueillies et publiées par Henri Dehérain, Paris, Société d'Éditions Géographiques, Maritimes et Coloniales, vol. I (1929) et vol. II (1930).

Thèses présentées à la faculté des sciences de Paris pour le Doctorat ès sciences physiques. Première thèse: Mémoire sur l'absorption de la chaleur rayonnante obscure dans les milieux de l'œil. Deuxième thèse: propositions de chimie données par la Faculté, Paris, Mallet-Bachelier, 1860.

Articles published in the following journals (a non-exhaustive list):

L'Aéronaute
Annalen der Physik und Chemie
Annales de chimie et de physique
Annales du Conservatoire national des arts et métiers
Annuaire du Bureau des longitudes
Annuaire du Club alpin français
Annuaire de la Société météorologique de France
Archives des missions scientifiques et littéraires
l'Astronomie
Atti della Pontificia Accademia Romana dei Nuovi Lincei
Bibliothèque universelle et revue suisse. Archives des sciences physiques et naturelles.
Bulletin astronomique
Bulletin de la Société française de photographie
Bulletin de la Société philomathique
Bulletin de la Société de géographie
Bulletin de la Société de météorologie
Chemical News
Conférence Scientia
Cosmos
Comptes rendus hebdomadaires de l'Académie des sciences
L'Institut
Journal d'hygiène
Memoirs of the Royal Astronomical Society
Memorie della Societa degli spettrocopisti Italiani

Les Mondes
Le Moniteur de la photographie
Le Moniteur du Puy-de-Dôme
Le Moniteur scientifique
Le Moniteur universel
Monthly Notices of the Royal Astronomical Society
La Nature
Nature
Philosophical Magazine
Proceedings of the Royal Society of London
Rapports de l'Association française pour l'avancement des sciences
Reports of the British Association for the Advancement of Sciences
Revue Scientifique
Vierteljahrschrift der Astronomischen Gesellschaft

Jules Janssen's Manuscripts

Archives de l'Académie des sciences de l'Institut de France (Paris): folders Janssen, Jean-Baptiste Dumas, & Lockyer; boxes Passages de Vénus *[Transits of Venus]*
Archives du musée Henner (Paris)
Archives du musée d'Orsay (Paris), Eiffel collection
Bibliothèque de l'Institut de France (Paris), notebooks and correspondance, Ms 4125 à Ms 4137.
Cambridge University Library, du Royal Greenwich Observatory archives.
Exeter University Library, Lockyer archives.
Société astronomique de France, Flammarion collection at Juvisy

Other Documents

Annuaire de La Marmite 1890, Paris, Imprimerie Richard.
500 célébrités contemporaines, Collection Félix Potin, Paris, 1904.
Discours prononcés aux funérailles de M. Janssen, Institut de France, Académie des sciences, Paris, Firmin-Didot, 1907.
Inauguration de la statue de Jules Janssen, Paris, Gauthier-Villars, 1920
ANONYME, « Janssen, de l'Institut », *Galerie des Notabilités contemporaines de la Revue d'Europe*, 1901.
AMALRIC Pierre, "Jules Janssen : from ophthalmology to astronomy", *Documenta ophtalmologica*, 81 (1992), 37–42.
AUBIN David, « La métamorphose des éclipses de Soleil », *La Recherche,* n° 321, juin 1999, 78–83.
AUBIN David, "Orchestrating observatory, laboratory, and field: Jules Janssen, the spectroscope and travel", *Nuncius*, **17** (2002), 615–633.
BADIN, Adolphe, « Madame Edmond Adam, Juliette Lamber », in *Étrennes aux Dames pour l'an 1882*, Paris, Charavay.
BAPPU M. K. V., *Astronomy in India during the period 1787–1947*, Ramakrishna Mission Lectures, 1974.
BEAU Mme Henri, née de Loménie, « Jules Janssen », in *Enfance des célébrités contemporaines, Revue L'Enfant*, Paris (1913), 23–29.

BECKER Barbara J., *Eclectism, opportunism, and the evolution of a new research agenda: William and Margaret Huggins and the origin of astrophysics*, Ph. D. dissertation, The Johns Hopkins University, 1993.

BECKER Barbara J., *Unravelling Starlight: William and Margaret Huggins and the Rise of the New Astronomy*, Cambridge (UK), Cambridge University Press, 2011.

BENNET Jim A., "The spectroscope's First Decade", *Bulletin of the Scientific Instrument Society*, 4 (1984), 3–6.

BIGOURDAN Guillaume, «Jules Janssen», *Bulletin astronomique*, **25** (1908), 49–58.

BRAUN Marta, *Picturing Time, The Work of Etienne-Jules Marey (1830–1904)*, Chicago and London, The University of Chicago Press, 1992.

BRAUN Marta, "The expanded present: photographing movement", in *Beauty of another order. Photography in Science*, Ann Thomas, ed., New Haven and London, Yale University Press, 1997, 150–184.

CANALES Jimena, "Photogenic Venus. The 'cinematographic turn' and its alternatives in the nineteenth-century France", *ISIS*, **93** (2002) 585–613.

CANALES Jimena, «Sensational differences: the case of the transit of Venus», *Cahiers François Viète*, **11–12** (2006), 15–40.

CAZIN Achille, *La spectroscopie*, Paris, Gauthier-Villars, 1878.

CHAM [NOÉ Amédée de], *Cours d'astronomie*, Paris, Plon, *ca* 1865.

CHAUVEAU Auguste, «Notice nécrologique sur M. Janssen», *CRAS*, **145** (23-12-1907), 1318–1319.

CLERKE Agnes M, *A popular history of astronomy during the nineteenth century,* Edinburgh, Adam and Charles Black, 1885.

COURAGE Sylvain [sous la direction de], «Spécial XIXe, le siècle de tous les possibles», *le nouvel Observateur*, n° 2250/2251, 20-12-2007, 67–114.

DAWES William, "Results of some recent observations of the solar surface, with remarks", *MNRAS*, **24** (1864), 161–165.

DÉBARBAT Suzanne, «Astronomes français et passage de Vénus sur le Soleil», *l'Astronomie*, **118** (2004), 268–273.

DÉBARBAT Suzanne et LAUNAY Françoise, "The objectives if the Great Paris Exhibition telescope of 1900", *Bulletin of the Scientific Instrument Society*, **74** (2002), 22–23.

DÉBARBAT Suzanne et LAUNAY Françoise, "The 1874 Transit of Venus observed in Japan by the French, and associated relics", *Journal of Astronomical History and Heritage*, **9** (2006), 167–171.

DE LA RUE Warren, "On a piece of apparatus for carrying out M. Janssen's method of time-photographic observations of the transit of Venus", *MNRAS*, **34** (1874), 347–353.

DESBEAUX Émile, *Physique populaire*, Paris, Flammarion, 1891.

DICK Steven J, ORCHISTON Wayne, and LOVE Tom, "Simon Newcomb, William Harkness and the nineteenth century American transit of Venus expeditions", *Journal for the History of Astronomy*, **29** (1998) 221–255.

DOLLFUS Audouin, «Le grand télescope de Janssen», *l'Astronomie*, **114** (2000), 236–250 et 360–371.

DOLLFUS Audouin, *La grande lunette de Meudon*, Paris, CNRS éditions, 2006.

DOLLFUS Audouin, «L'observation des étoiles filantes Léonides en ballon de 1898 à 1900», *l'Astronomie*, **121** (2007), 568–572.

DUMONT Simone, «Les expéditions du XVIIIe siècle pour les passages de Vénus», *l'Astronomie*, **118** (2004), 290–295.

DURAND-GRÉVILLE Émile, *Entretiens de J. - J. Henner*, Paris, Lemerre, 1925.

DURIER Charles, *Le Mont-Blanc*, Paris, Fischbacher, 4ème édition, 1897.

FAIDIT Jean-Michel, «La fête du Soleil», *l'Astronomie*, **118** (2004), 370–373.

FAUQUE Danièle, «Observatoires astronomiques français et diffusion de l'astronomie à l'Association française pour l'avancement des sciences (1872–1914)», in *Observatoires et patrimoine astronomique français*, G. Boistel dir., Lyon, SFHST/ENS-LSHE Éditions, 2005, 175–189.

FÉTIS François Joseph, *Biographie universelle des musiciens*, Paris, 1874.

FLAMMARION Camille, «Le passage de Vénus. Résultats des expéditions françaises», *La Nature*, n° 101, 8 mai 1875, 356–358.

FLAMMARION Camille, «L'astronome Janssen», *Bulletin de la Société astronomique de France*, **22**(1908), 25–28.

FONVIELLE Wilfrid de, «M. Janssen au Mont-Blanc», *Journal des Voyages*, n° 234, 26 mai 1901, 410–411.

FRAUNHOFER, Joseph, "Bestimmung des Brechungs - und Farbenzerstreuungs - Vermögens verschiedener Glasarten in Bezug auf die Vervollkommnung achromatischer Fernröhre", *Denkschriften der königlichen Akademie der Wissenschaften zu München*, **5** (1817), 193–226.

GAPAILLARD Jacques, *Une histoire de l'heure en France*, Paris, Vuibert, (in press, 2011)

GEORGELIN Yvon et et ARZANO Simone, «L'éclipse de Soleil du 18 août 1868, Stephan et Rayet, hôtes du roi de Siam à Wha-Thonne», *l'Astronomie*, **113** (1999), 12–17.

GUILLEMIN Amédée, *Le Ciel*, Paris, Hachette, 1877.

HARDING, Colin, "Images on disc", *Photographica World*, n°62, September 1992, 7–12.

HEARNSHAW John, "Photographic stellar photometry in the late 19th and early 20th centuries", in *100 Years of Observational Astronomy and Astrophysics, Homage to Miklos Konkoly Thege (1842–1916)*, C. Sterken and J. Hearnshaw ed., 2001, 171–182.

HERMAN Jan K. and CORBIN Brenda G., "Trouvelot: From Moths to Mars", *Sky & Telescope*, Dec. 1986, 566–568.

HINGLEY Peter D., "Thomas Skaife's pistolgraphs", *Astronomy and Geophysics*, **40** (1999), 5.7.

HINGLEY Peter et LAUNAY Françoise, «L'éclipse de 1860», in *Dans le champ des étoiles, les photographes et le ciel 1850–2000*, RMN, Paris (2000), 117–118.

HINGLEY Peter et LAUNAY Françoise, «Passages de Vénus, 1874 et 1882», in *Dans le champ des étoiles, les photographes et le ciel 1850–2000*, RMN, Paris (2000), 119–120.

HORN D'ARTURO Guido, "Il fenomeno della 'goccia nera' e l'astigmatismo", *Pubblicazioni dell'Oservatorio astronomico della R. Università di Bologna*, **I n° 3** (1922), 25–57.

HOWSE Derek, *Greenwich time: and the discovery of the longitude*, Oxford, Oxford University Press, 1980

JANSSEN Mme, «Souvenirs d'un voyage aux Nilgherries, les Todas», *Le Tour du Monde*, **XLIV** (1882), 1137e livraison, 241–249.

KNOBEL E. B., "Pierre Jules César Janssen", *Monthly Notices of the RAS*, **68** (1908), 245–249.

KUIPER M. (ed.), *The solar system. I The Sun*, Chicago, The University of Chicago Press, 1953.

LA BAUME PLUVINEL Aymar de, "Jules César Janssen", *The Astrophysical Journal*, **28** (1908), 88–99.

LA COTARDIÈRE Philippe de, et FUENTES Patrick, *Camille Flammarion*, Paris, Flammarion, 1994.

LAFLEUR Martine, «Autour de la controverse soulevée par la mise à l'index de quatre manuels scolaires français de la fin du XIXe siècle : l'appartenance sociale à la République française et la politisation de la différence des sexes», *Mémoire de Maîtrise ès Arts*, Université de Sherbrooke, Canada, 1998.

LANCE Adolphe, *Dictionnaire des architectes français*, Paris, 1872.

LAPPARENT Albert de, «Les deuils de la science française. Janssen», in *Le Correspondant*, Paris, L. de Soye et fils, 1908.

LAUNAY Françoise, «Autour de l'éclipse totale de soleil visible en Inde le 12 décembre 1871 : le voyage du couple Janssen aux Neelgherries», *l'Astronomie*, **111** (1997), 111–115.

LAUNAY Françoise, «Jules Janssen et la photographie», in *Dans le champ des étoiles ; les photographes et le ciel 1858–2000*, RMN, Paris (2000), 22–31.

LAUNAY Françoise, «Le Soleil», in *Dans le champ des étoiles, les photographes et le ciel 1850–2000*, RMN, Paris (2000), 127.

LAUNAY Françoise : «Spectroscopie : Soleil et étoiles», in *Dans le champ des étoiles, les photographes et le ciel 1850–2000*, RMN, Paris (2000), 134–135.

LAUNAY Françoise, "Jules Janssen's in and out correspondence", *100 years of observational astronomy and astrophysics, Homage to Miklos Konkoly Thege (1842–1916)*, C. Sterken and J. Hearnshaw ed., Brussels (2001), 159–168.

LAUNAY Françoise, «Trouvelot à Meudon: une "affaire" et huit pastels», *l'Astronomie*, **117** (2003), 452–461.

LAUNAY Françoise, «Le revolver photographique de Jules Janssen», *l'Astronomie*, **118** (2004), 306–308.

LAUNAY Françoise, «Jules Janssen et la création de l'observatoire de Meudon (1875–1879)», *Bulletin du Comité de Sauvegarde des Sites de Meudon*, **119–120** (2006), 22–28.

LAUNAY Françoise «Les acteurs de la mission de Jules Janssen au Japon en 1874», *l'Astronomie*, **121** (2007), 94–101.

LAUNAY Françoise, «La statue de Jules Janssen (1824–1907) sur la terrasse publique de Meudon», *Bulletin des Amis de Meudon*, n° 237, avril 2007, 23–29.

LAUNAY Françoise, "The great Paris exhibition telescope of 1900", *Journal for the History of Astronomy*, **38** (2007), 459–475.

LAUNAY Françoise, «La correspondance Janssen/Flammarion», *l'Astronomie*, **121** (2007), 558–566.

LAUNAY Françoise and HINGLEY Peter D., "Jules Janssen's 'Revolver photographique' and its British derivative: the 'Janssen slide'", *Journal for the History of Astronomy*, **36** (2005), 57–79.

LAUNAY Françoise and HINGLEY Peter D., "Astronomy", in, *Encyclopedia of the Nilgiri Hills*, Paul Hockings (ed.), New Delhi, Mahonar Books, 84–90, to be published.

L[AURENCIN] P., «Les instruments de la station astronomique française du Japon», *L'Illustration*, n° 1639, 25-07-1874, 55 et 64.

LECLERC Joseph, «Jules Janssen et le cinématographe», *l'Astronomie*, **69** (1955), 388–391.

LE COCGUEN Régis et LAUNAY Françoise, «La lunette de photographie solaire de Janssen», *l'Astronomie*, **119** (2005), 71–75.

LE GARS Stéphane, «La spectroscopie solaire: les contributions de Jules Janssen», mémoire de DEA, Université de Nantes, 2002.

LE GARS Stéphane et MAISON Laetitia, «Janssen, Rayet, Cornu: trois parcours exemplaires dans la construction de l'astronomie physique en France (1860–1890)», *Revue d'histoire des sciences*, **59** (2006), 51–81.

LE GUET-TULLY Françoise, «La création de l'observatoire d'Alger», *La Revue*, n° 38 (juin 2003), 26–35.

LÉVY Jacques, «Janssen, Pierre Jules César», in *Dictionary of scientific biography*, Charles C. Gillispie ed., New York, Charles Scribner's Sons, 1973, Vol. 7, 73–78.

LOCKYER, J. Norman, *Contributions to Solar Physics*, London, Macmillan and co, 1874.

LOCKYER J. Norman, «The story of Helium», *Nature*, **53** (1896), 319–322 et 342–346.

LOCKYER Norman, "Obituary Notices of Fellows deceased, P. J. C. Janssen, 1824-1907", *Proceedings of the Royal Society of London*, **81** (1908–1909), 77–81.

LOCKYER T. Mary and LOCKYER Winifred I., *Life and Work of Sir Norman Lockyer*, London, Macmillan, 1928.

MACPHERSON Hector, Junior, "Pierre Jules César Janssen", in *Astronomers of to-day*, London and Edinburgh, Gall and Inglis, 1905.

MAHIAS Marie-Claude, «Le Soleil noir des Nilgiri, l'astronomie, la photographie et l'anthropologie physique en Inde du sud», *Gradhiva*, **24** (1998), 45–56.

MALHERBE Jean-Marie, «Les observatoires du mont Blanc en 1900», *l'Astronomie*, **101** (1987), 115–134.

MALHERBE Jean-Marie, «Janssen et l'Observatoire astronomique du mont Blanc», *l'Astronomie*, **107** (1993), 278–283.

MANNONI Laurent, *Trois siècles de cinéma*, Paris, Éditions de la Réunion des musées nationaux, 1995.

MARLOT Christophe, *Les passages de Vénus*, Paris, Vuibert, 2004.

MEADOWS Arthur J., *Science and Controversy. A biography of Sir Norman Lockyer*, Macmillan, 1972.

MÉRY Gaston, in *La libre Parole*, 24–05, 27–05, 29–05 et 06-06-1895.

MILLOCHAU Gaston, «La grande lunette de l'Observatoire du Mont-Blanc», *La Nature*, 1906, 23–26.

Milogradov-Turin Jelena, "Origin of the photospherical network - The main astrophysical research of Đorđe Stanojević", *Publi. Astron. Obs. Belgrade*, **80** (2006), 265–268.

Nodon Albert, «Portraits de savants. Jules Janssen (1824–1907)» in *La Science moderne*, juillet 1926, 387–392.

Nordmann Charles, «Jules Janssen», *Revue générale des sciences*, **19** (1908), 223–225.

Pasachoff Jay M., Olson Roberta J. M., and Hazen Martha L, "The earliest comet photographs: Usherwood, Bond, and Donati 1858", *Journal for the History of Astronomy*, **27** (1996), 129–145.

Pasachoff, Jay M., Schneider, Glenn and Golub, Leon, "The black-drop effect explained," in *Transits of Venus: New Views of the Solar System and Galaxy*, IAU Colloquium No. 196 (U.K., 2004), D. W. Kurtz and G. E. Bromage, eds., 2005, 242–253.

Pasachoff, Jay M., and Suer, Terry-Ann, "The Origin and Diffusion of the H and K Notation," *Journal of Astronomical History and Heritage*, **13–2** (2010), 121–127.

Pector S, «J. Janssen», *Bulletin de la Société française de photographie*, **24** (1908), 76–78.

Richalet Jean-Paul, "The scientific Observatories on Mont Blanc", High altitude medicine and biology, **2** (2001), 57–68.

Roberts Dorothea [née Klumpke], "Pierre Jules César Janssen, (1824–1907)", *Bulletin de l'observatoire populaire de Rouen*, 1908, 47–60.

Rösch Jean, «Photographies de la photosphère et des taches solaires», *l'Astronomie*, 71 (1957), 129–141.

Salet Pierre, *Spectroscopie astronomique*, Paris, Doin, 1909.

Secchi Angelo, Le Soleil, 2e édition, Paris, Gauthier-Villars, 1875.

Sheehan, William and Westfall, John, *The Transits of Venus*, Amherst, NY, Prometheus Books, 2004.

Sicard Monique, «Passage de Vénus. Le Revolver photographique de Jules Janssen», *Études photographiques*, **4** (1998), 44–63.

Tarbell Ida M., "The Observatory on top of Mt Blanc", *Mc Clure's Magazine*, February 1894, 289–301.

Thomas André, *La Société philomathique de Paris et deux siècles d'histoire de la science en France*, Paris, PUF, 1990.

Thomine Alice, *Émile Vaudremer 1829–1914. La rigueur de l'architecture publique*, Paris, Picard, 2004.

Tikhov Gabriel, *L'énigme des planètes*, Union soviétique, Éditions en langues étrangères.

Tissandier Gaston, «Photographies de nos contemporains chez eux. M. Janssen», *La Nature*, n° 1168 du 19 octobre 1895, 329–331.

Tissandier Gaston, «L'Observatoire du mont Blanc», *La Nature*, n° 1062, 07-10-1893, 289–290.

Tosi Virgilio, «Les origines du cinéma scientifique 1», Paris, CNRS audiovisuel, 1997.

Touchatout (Bienvenu Léon Charles), *Le trombinoscope* n° 62, Paris, 1873.

Toulmonde Michel, «La parallaxe du Soleil», *l'Astronomie*, **118** (2004), 274–289.

Toulmonde Michel, *Les rendez-vous de Vénus*, CD-Rom, Observatoire de Paris, 2004.

Tournaire R., «Jules Janssen, astronome français (1824–1907)», *Les contemporains*, n° 1049, 17-11-1912.

Ueberschlag Josette, «Janssen et le démon de la précision», *l'Astronomie*, **114** (2000), 58–61.

Véron Philippe, *Dictionnaire des astronomes français, 1850–1950*, unpublished.

Villadier, Francis et Marie-José, *L'aérostation à Meudon*, Ville de Meudon, 1983.

Vivian Robert, *L'épopée Vallot au mont Blanc*, Paris, Denoël, 1986.

Vivian Robert, *L'épopée Vallot au mont Blanc: 100 ans déjà...*, Montmélian, La Fontaine de Siloé, 1998.

Warner Deborah Jean, "Direct vision spectroscopes", *Rittenhouse,* **7** (1993), 40–48.

Yvert Benoît [sous la direction de], *Dictionnaire des ministres (1786–1989)*, Paris, Perrin, 1990.

Name Index

(When known, only the usual first names are given)

MIX
Papier aus verantwortungsvollen Quellen
Paper from responsible sources
FSC® C105338

If you have any concerns about our products,
you can contact us on
ProductSafety@springernature.com

In case Publisher is established outside the EU,
the EU authorized representative is:
Springer Nature Customer Service Center GmbH
Europaplatz 3, 69115 Heidelberg, Germany

Printed by Libri Plureos GmbH
in Hamburg, Germany